COMPLIANCE
MADE EASY

What The Top Cyber Security Specialists Do To
Accelerate HIPAA Certification

ISBN: 979-8-9862097-4-6
LCCN: 2022923007

Most TechnologyPress™ titles are available at special quantity discounts for bulk purchases for sales promotions, premiums, fundraising, and educational use. Special versions or book excerpts can also be created to fit specific needs.

For more information, please write:

TechnologyPress™
3415 W. Lake Mary Blvd. #950370
Lake Mary, FL 32746
or call 1.877.261.4930

COMPLIANCE
MADE **EASY**

What The Top Cyber Security Specialists Do To
Accelerate HIPAA Certification

TechnologyPress™
Lake Mary, Florida

CONTENTS

CHAPTER 1

THE CASE FOR CYBER-COMPLIANCE

BY PAUL TRACEY,
Owner, Innovative Technologies.

While there's no fail-safe method to secure data, we know for certain that cybercriminals are too smart to be deterred by bare-minimum cyber security measures. Truth is, most business leaders do not implement sufficient safety protocols, much less consider how their negligence affects others. And small organizations may not even understand the severity of potential collateral damage from a breach.

Cyber security is messy. It's complex. Frankly, most organizations don't do a great job when left to their own devices. Fortunately, though, governments step in to protect the public with strict cyber-compliance regulations for certain industries, with steep penalties for violations.

THE SNOWBALL EFFECT OF GETTING HACKED

Cyber-compliance is necessary because the consequences of a data breach extend far beyond an organization's own bottom line. First, if the attack interrupts operations and requires a shutdown – no matter how temporary – this puts those who rely

11

on its services or products in a bind. At a minimum, it will affect the local economy. If you run a health care practice, patients will scramble to find a new provider or suffer without treatment. And if you sell N95 respirators and suspend manufacturing and delivery, you'll disrupt the supply chain, giving doctors and nurses no choice but to re-wear personal protective equipment that may no longer be effective.

These hypothetical scenarios are based on real-life disasters such as the massive ransomware attack on Colonial Pipeline, the east coast's primary fuel distributor. When they were hit, to temper further risk, they froze deliveries. As a result, gas stations suffered financially, and the supply shortage caused long lines, holding up small businesses that needed fuel to transport goods and passengers. Employees struggled to fill tanks to travel to work, and service industry professionals such as stylists, dog walkers and contractors had to cancel jobs, leading to lost income and damaged reputations.

This hack occurred because the company, in its failure to follow any basic best practices, exposed a VPN account password, which would be a violation of any set of compliance rules. In this case, a single careless act had profound effects on society.

Second, when your organization is attacked, hackers now have easier access to everyone you serve or who serves your business, posing a higher threat level for anyone with data on the system – both local and cloud-based. The hacker will immediately go after your business and personal contacts. If you run a medical practice, think about all the patients, customers and associates whose privacy would be compromised if your noncompliance led to a breach.

And imagine the fallout if an associate receives an e-mail that looks like it came from you, so they open it, but it contains ransomware. Or if they open a message from a cybercriminal that says, "We have your data because we hacked (insert your practice's name). If you want it, you will have to buy it back from us."

In the reverse, if your clients or associates experience breaches, auditors will validate the data that was transferred when engaging with your network. To protect your organization, always maintain a clean bill of compliance. Otherwise your organization could face potential civil and/or criminal liability, depending on the outcome of the incident. Also, be aware that even if the data did not originate from your business, you may still face compliance fines.

HOW DO CYBER SECURITY AND COMPLIANCE DIFFER?

Anyone running a business must understand the lines between cyber security and cyber-compliance, and the fact that security measures alone are not sufficient. As leaders, it is incumbent upon us all to honor our social responsibility to follow cyber-compliance regulations so we can support the safety of those whose lives we impact.

The differences between levels of protections are similar to those that apply to automobile safety. Auto manufacturers are required to build the vehicles that we drive with seat belts, airbags, a horn, directionals, headlights, brakes, brake lights and door locks. These features aid in the operation of the vehicle, protect us from harm in the event of a crash, alert others about our movements and prevent thieves from easily entering. A truck or tractor trailer with bad brakes, for example, not only puts its driver and passengers at risk, but this malfunction increases the likelihood that it will crash into another vehicle or even a pedestrian.

Auto safety features are akin to the basic levels of cyber security needed for computer networks. This includes antivirus software to aid in the machine's defense capabilities, firewalls to regulate incoming and outgoing network traffic, phishing training so users know how to perceive and avoid danger, and a maintenance program that ensures all software patches are up-to-date.

Compliance, though? That's the powerful stuff. These are the laws governments enforce based on agreements we have made as a society about safety. Rules of the road are based on the consensus that we do not want drivers to be under the influence of alcohol or drugs, weave in and out of traffic or exceed certain speeds based on terrain and weather conditions, and that mechanical safety features must be in working condition and used properly.

And although most of us don't understand the technical aspects of cyber security, we agree that we don't want reckless cyber behavior to put others at risk. Cyber security compliance regulations are in place to set guidelines for actions needed to take care of each other.

WHY DO SOME INDUSTRIES HAVE COMPLIANCE LAWS AND OTHERS HAVE NONE?

The rationale for inconsistent compliance legislation is that the financial impact of breaches varies by industry. Health care data is worth more money to criminals than all other types – even information held by financial services companies – so providers and other covered entities are held to the compliance regulations dictated by the Health Insurance Portability and Accountability Act of 1996 (HIPAA). And due to the snowball effect discussed above, business associates must also comply with the same level of regulations as the health care vendor.

Besides the monetary value that makes certain data more appealing to hackers, the level of risk is another factor that determines the strictness of cyber security laws. It's the same as with transportation. If you drive a car for personal use only, no one is concerned about your service history or how many hours you've been behind the wheel during a given time period. However, a commercial driver operating an 18-wheeler must use the truck and its security features in a manner defined by industry regulations and produce the documentation to prove compliance.

Trucks that transport hazardous materials are accountable to additional rules, just like higher-risk verticals such as finance and defense contracting are subject to a stricter set of cyber security requirements than others. Federal and State legislatures determine the standards to which each industry is held as well as the urgency of implementing compliance laws.

What we find, unfortunately, is that the formula – based on the market value of the data – often backfires. This is because hacking into a system that contains less lucrative data provides a path to that which holds a higher value. Once inside, hackers can reach the vendor's network and quietly stay dormant while their ransomware spreads to all contacts. The smaller companies with less mature security practices are often access points to the data of larger, more secure organizations. One such third-party attack happened when Target was hacked after allowing a heating, ventilation, and air-conditioning technician to connect to their network using an infected machine. This exposed seventy million shoppers' data.

KEEPING UP WITH US AND EU REGULATIONS

One of the US government's latest regulatory moves in cyber security is forced reporting. A prime example is the Cyber Incident Reporting for Critical Infrastructure Act, which passed in March of 2022. This requires covered entities (CEs) in health care to report cyber-attacks and ransomware payments to the Cybersecurity and Infrastructure Security Agency. The intent is to make it easier for CISA to act fast to mitigate the effects of attacks in health care, support victims, analyze trends and disseminate information to those who benefit from it. While data compromises happen quickly, businesses had typically delayed reporting, taking 30 to 60 days to notify those affected, which allows for more damage to occur.

US companies can expect the scope and number of reporting laws to increase dramatically through 2025. If your organization has a breach, be prepared to follow shorter time limits and new regulations around disclosures.

In 2018, the European Union introduced the General Data Regulation Protection privacy and security law, the strictest set of cyber-rules on the planet. GDRP even dictates the protection and accountability practices for any organization around the world that accesses data from persons in the EU.

HOW COMPLIANCE IMPACTS CYBER INSURANCE COVERAGE

Everyone wants to think their cyber insurance will pay out for damages lost due to a hack, but this is only possible when an organization followed policy compliance and can validate it. Again, it comes back to agreements. The insurance company expects customers to follow the terms of the contract in order for them to do their part.

If you've ever applied for cyber insurance, you know it is an arduous process that involves answering up to 300 questions and providing a full report of all data and locations, as well as proof that two-factor authentication is set up on all user accounts. The insurance company requires this extensive questionnaire and documentation to minimize the chances that the insured will collect on the policy.

Similarly, if a driver causes an accident, stakes are highest for commercial truck drivers who are required by insurers to comply with the Department of Transportation rules for documentation. If the truck crashes, and there's no evidence of brake inspections in six months and the worn-down pads are determined as the cause of the accident, insurance will not pay.

Six months is also too long to go without updating antivirus protections, and insurance compliance will tell you this. If your network is the victim of a cyber-attack or ransomware and you lose all your data, but security products aren't up-to-date, insurance will not pay out.

Frequently, people looking to renew their policy call my company, a Managed Security Services Provider (MSSP), in a panic, asking us to quickly implement security measures like two-factor authentication. Tools deployed in a hurry are rarely set up properly, though, and don't follow procedures and policies that satisfy compliance, which can bring your business to a screeching halt. We're happy to help with your insurance application but expect it to take time. The best option is to prepare for these changes when you are not rushed – long before your renewal.

Once your cyber insurance policy is approved, continue to create documentation to certify that you're doing the right things on an ongoing basis. These records are a large part of compliance.

REGULAR MAINTENANCE – A NECESSARY COMPONENT OF COMPLIANCE

Too often, we find organizations don't have a maintenance plan in place – and don't even know they need one. C-suite executives need to work with IT to manage and mandate network maintenance through policy and not defer to the IT department. Taking responsibility for regularly reviewing security policies or network health ensures that IT is taking all necessary steps to inspect and fix the network issues as they arise.

The same collaborative relationship is essential between drivers and mechanics. Truck accidents are generally caused by bad maintenance, broken equipment, or hardware failures that the driver wasn't aware of until facing an OMG moment. This happens when they're not getting inspected frequently enough. If the mechanic says your truck's alignment looks great, but you neglect maintenance, it's not their fault if, months later, your brakes fail and you crash.

Cyber-compliance also requires involving the entire team and keeping everyone up-to-date with annual refresher training. The danger with complacency in tech is that the world is different today than when we went to sleep last night. The fix a cyber security pro did yesterday that stopped threat X could open up threat Y today. And the content from that IT security course your team took five years ago is definitely obsolete. Be sure to build continuing education into your company's culture.

Additionally, do more than required. In my opinion, HIPAA is too relaxed on checkups. It only mandates a security risk assessment once a year. The administrators say they'd "like you to do it every quarter." That is not sufficient. You need to do this on a continual basis, since the data you hold changes frequently. If you have a medical practice, new patient information is likely entered into the system daily, and when anyone comes in for an appointment and is prescribed treatments and medications, more new data is produced for all services.

You are responsible for that data at rest and in transit, therefore security measures need to be documented for compliance. Continually check how data is coming into the system, where it is stored and that it's all being backed up in real time. Neglecting this last step can be fatal for newer organizations with lower revenue, like one small business that got attacked recently. The owner told me his provider tried to sell him the better backup solution but wasn't insistent enough, so he opted out. He didn't have the protections in place to stay afloat after getting hacked and he had to shut down.

WHY IS COMPLIANCE THE ANSWER?

It's nearly impossible to fathom the amount of data each of us, as individuals, have online. And businesses likely communicate with every vendor via e-mail, and maybe through an online billing system too. All of that data has to flow, and if at any point they are not secure, both parties at the ends of that transaction become vulnerable.

Most business owners don't understand the technical side of risks – how the threat works and how data could be stolen – and that's okay. Compliance laws in particular are created by people who do understand how data leaks happen, to outline expectations for those held to the law. The important part is to know your responsibilities.

HOW CAN YOU ENSURE COMPLIANCE?

To be HIPAA compliant or adopt this level of security measures in any industry, you need a comprehensive security and compliance program. You have to test against your policies, identify your security gaps, make a plan to fix them, and test and confirm whether that security issue has been remediated. These are requirements for both HIPAA and the Cybersecurity Maturity Model Certification compliance. CMMC is the HIPAA equivalent for federal government contracts. The two sets of requirements have many similarities, and we expect to see some crossover in policy adoption.

HIPAA compliance also requires six specific audits/assessments per year. However, almost no one is doing all those audits or even understands the differences between them. This is a good indicator of why security and compliance are not the same thing.

Another factor that's becoming more important is the maturity of your security, so it's crucial to button up cyber security measures not only often – but early.

HOW TO KNOW YOUR MSSP HAS YOUR BACK

Beware of any MSSP who will sell you less than you need, then convince you over time to implement additional protocols. Our MSSP insists on a minimum level of service to ensure you get not only what you want, but what you need to satisfy regulations your business faces. Anyone with access to your systems should meet state regulations and other compliance requirements. Be sure to work with an MSSP that cares about your safety and that of others you affect. Compliance can not only protect your organization but also help you maintain status as a good digital citizen that plays well with others.

Sources:
- https://www.pcmag.com/news/hvac-vendor-confirms-link-to-target-data-breach
- https://www.cisa.gov/circia
- https://www.hhs.gov/hipaa/for-professionals/covered-entities/index.html
- https://cybersecurity.att.com/blogs/security-essentials/what-is-cmmc-compliance

About Paul

Compliance expert Paul Tracey lives, breathes, writes, and speaks about taking all possible measures to lock down an organization's data. He will tell you that every business should follow HIPAA requirements – the most stringent of all compliance regulations – and even step up those standards with more frequent system audits.

Paul believes making a high level of security mandatory will better protect those who suffer when their data is on a system that is breached. While working a decade ago at a large health care institution that was hacked, he first started to pay attention to the fact that many sizable organizations do not take cyber security seriously. He then noticed bigger companies would often brush off a breach – pay the fines with a shrug and be unscathed by any further consequences.

To provide affordable technology services to small businesses that may not otherwise survive a hack, Paul started his company, Innovative Technologies. IT industry experience, as well as studies in information technologies and business at SUNY Adirondack, gave him the tools to minimize security risks for small and medium-sized businesses nationwide.

Today, Innovative Technologies continues to help clients ensure they have security and compliance procedures in place and a well-trained staff. Their approach and expertise has earned them their reputation as a leading Managed Security Services Provider. Keeping up with training is a value Paul espouses. He renews his HIPAA Seal of Compliance Verification annually and leads the company's compliance program.

Paul is a recognized cyber security expert who has appeared on national and New York–based news segments on ABC, CBS, Fox, and NBC, along with several podcasts. He is an Amazon bestselling author and writer of the book *Cyber Storm*, and his writings have been published by more than 500 news outlets, including the *Boston Herald*, which carried his piece "Staying Cyber Secure While Working From Home." Paul has also been featured in *MSP Cyber Security* and *MSP Success* magazines and is the author of *Delete the Hacker Playbook*.

Outside of work, Paul has served as chief technology officer and board member for the Glens Falls Greenjackets semi-pro football team. Innovative Technologies sponsors the team and regularly donates to St. Jude Children's Research Hospital.

Profound honesty, transparency and humility are values that permeate Paul's work. And the sign on his office wall, "Never Stop Auditing," applies beyond the literal to his practice of perpetually assessing situations and seeking new solutions.

Innovative Technologies
- Address: Malta, NY (Saratoga County)
- Website: www.upstatetechsupport.com
- Email: info@upstatetechsupport.com
- Tel: (518) 900-7004

CHAPTER 2

WHY COMPLIANCE AS A CULTURE MUST BE TAKEN SERIOUSLY FROM THE TOP DOWN

BY WILLIAM DICKHERBER,
Founder, Onsite Computer Consulting Services

Consider two scenarios:

- In a doctor's office, a patient engages the person at the front desk in a discussion regarding their account. It escalates, with the employee demanding payment for an overdue account in front of other patients and families in the waiting room.
- A dentist's patient posts an unflattering online review of their visit, writing that they didn't like the way they were treated. Someone who works for the dentist responds, "Pulling teeth hurts."

In the first case, a patient in the waiting room was witness to the altercation at the front desk that not only exposed the balance of the patient's account, but also the fact that it was past due. In the second, the employee, wittingly or unwittingly, broadcast to the world that this particular patient had to have a tooth pulled.

Both cases are violations of the Health Insurance Portability and Accountability Act (HIPAA), a federal law passed in 1995 that "required the creation of national standards to protect sensitive patient health information from being disclosed without the patient's consent or knowledge," according to the CDC.

To the layperson, both may seem to be benign incidents, but medical practitioners and health care service providers mishandle them at their own peril. Compliance entails many areas within the business, including cyber security, client interaction skills, accounting, and inventory guidelines. Compliance is also a term that incorporates any business or entity that deals with personal client data, including medical, financial, or personal information.

HIPAA violations can have serious consequences that range from loss of money and reputation to, in extreme cases, the loss of one's business. Believe me, I have seen it happen. Committing to best compliance practices needs to be embraced at all levels, by all employees.

Before we can talk about establishing a top-down culture of HIPAA compliance, we first need to understand HIPAA transaction and code sets standards, which were created in part to ease administrative burdens on such entities as health care providers, health care plans, health care clearinghouses and business associates.

Health care providers, regardless of practice size, electronically transmit health information in connection with certain transactions that include claims, benefit eligibility inquiries, referral authorization requests and other transactions for which the US Department of Health and Human Services (HHS) established standards.

Health care plans are established by entities to pay the costs of medical care, including health, dental, vision and prescription drugs. They also include employer-sponsored group health plans, government- and church-sponsored health care plans and multiemployer health care plans. (A group health care plan with fewer than 50 participants that is administered solely by the employer that established and maintains the plan is not a covered entity.)

Health care clearinghouses, in most cases, involve entities that process nonstandard information they receive from another entity into a standard format or data content. They receive individually identifiable health information only when they are providing these processing services to a health care plan or health care provider as a business associate.

Business associates are people or organizations using or disclosing individually identifiable health information to perform or provide functions, activities, or services for a covered entity. These functions, activities or services include claims processing, data analysis, utilization review and billing.

It is important to underscore that it is not just health care offices, but also the businesses that support the health care industry, that are directly affected by HIPAA requirements. According to *HIPAA Journal*, there were an average of 59 data breaches each month in 2021. There were 712 health care data breaches reported between January 1 and December 3, a roughly 11% increase from 2020. Taking compliance seriously promotes security within your organization.

What comprises an organizational culture of compliance? Ask yourself these questions:

- Does your business's culture incorporate a commitment to compliance principles?
- Does the culture allow the resources that support the compliance principles?
- Is the culture aware of individual achievement and does it allow for the recognition of and reward for compliant behavior?
- Does the culture hold individuals and the group accountable and provide consequences for identified violations?

- Is your culture such that your employees feel inspired to act compliantly and are encouraged to feel safe reporting violations?

A positive culture of compliance provides employees and clients a safe and secure experience.

When HIPAA compliance was introduced, many of those at the top levels of companies identified it as a bureaucratic necessity, but I noticed with smaller companies, especially, a laissez-faire attitude toward HIPAA compliance. The original HIPAA Policies and Procedural Manual from 2015 would be gathering dust on the shelf, unread.

Why would this be? The most common reasons business owners gave me were "I need to focus on leading the company" and "It's somebody else's problem." Employees tend to take their cues from upper management. Between a lack of training and involvement, they embrace the same indifference toward compliance culture and data security. This leaves companies at risk of being assessed prohibitive fines for violations.

Horror stories abound. In 2018, Anthem Inc. settled a HIPAA violation case stemming from a 2015 data breach that involved protected health information of nearly 80 million plan members – at $16 million, it was the largest HIPAA settlement to date.

Leadership must lead by example and take a primary role in the implementation and integration of a compliance culture. Assigning a HIPAA officer in the organization and delegating duties to various individuals within the organization does *not* relieve top management of the final responsibility of compliance. The old saying "The buck stops here" is particularly true in this situation.

I impress upon business leaders that they are actually building a culture inside their own realm and creating their own business of HIPAA. Implementing common-sense best practices will protect your employees and your business. It is imperative that your staff buy into the compliance processes and that they are made aware of the rules. While some employees willfully violate HIPAA (like the nineteen UCLA Medical Center employees who were terminated after they took it upon themselves to access Britney Spears's medical records following her 2008 psychiatric hospitalization), many HIPAA violations are due to poor training.

As mentioned above, social media is a common breeding ground for unwitting HIPAA violations. In 2010, a nurse was terminated after she posted about a patient with a gunshot wound and who was accused of killing a policeman. In 2017, a med. technician was fired after she posted on Facebook, about a car-crash victim, "Should have worn her seat belt."

Compliance responsibility does not end after business hours. Filefax, a firm based in Northbrook, Illinois, that provided for the storage, maintenance, and delivery of medical records for covered entities was assessed a fine of $100,000 after a dumpster-diver retrieved the medical records of approximately 2,150 patients; these records had not been properly disposed of and the dumpster-diver sold the paperwork. The company ceased operations during the investigation. It is no longer in business.

So, staff needs to be taught: "You can't do this because you're going to wind up unemployed."

The leadership of the organization that produces a positive culture of compliance finds that the overall environment provides for less conflict and a more positive customer and employee interaction. HIPAA regulations clearly stipulate actions, responses and resolutions for most issues encountered in daily operations. Employees often think they are representing the company's best interests when confronting a client over issues at the front desk. But, as mentioned above, clashes with a patient, if they occur in front of other patients, are a direct HIPAA violation. This is significant to acknowledge – it does not have to be the individual customer who was directly involved who can notify HHS, the governing body for HIPAA, of a violation. It could be anyone present at the time or even a third party who "heard" about the incident. Keep this in mind when determining your involvement with training and situational awareness around compliance.

One sensitive part of establishing a positive culture of compliance is dispelling fears of reprisal for pointing out areas of compliance that are not adhered to and citing small issues for correction before they become violations. Employees will be less resistant to doing this in a positive, top-down culture of compliance that communicates a clear message that noncompliant, unethical, illegal and/or immoral behavior will not be tolerated. Employees recognizing that top management is fully engaged with compliance also know that recognition, reward, and positive implementation, and not repercussions, would be more the norm.

For a company to grow, it needs to implement a rigorous compliance program that integrates training and enhancement of policies that best reflect the mission statement and policies of the company. This will enhance the very essence of the company, building the team approach and creating a positive operating environment.

I cannot stress enough that a culture of compliance starts at the top. It is virtually impossible to guide someone on the path of compliance if you yourself are unsure of what the policies and procedures are. Be aware that HIPAA compliance is ever-changing as new threats to data protection emerge. Training your compliance officer to recognize and be situationally aware of potential issues in your organization is imperative.

A top-down culture of compliance teaches employees the things they need to know to keep the company and each other safe from bad actors while providing them the enthusiasm to learn and accept the compliance responsibility that is required. When executives prioritize, promote, and even participate in training, everyone gets the message that it's important.

The benefits of a culture of compliance are several-fold:

1. When an employee is trained through interactive, relevant training scenarios, they can see the direct impact of a scenario and how it can positively or negatively impact the company. This provides real-life experience to be used in day-to-day situations.
2. Through repeated training and situational role-playing, the employee achieves a practiced response to situations.
3. Compliance-related guidelines and best practices are fluid. Ongoing training will keep employees up-to-date as new HIPAA requirements are identified.

When it comes to HIPAA compliance, the most common complaint I hear from business owners and employees is that, as with many government-directed programs, HIPAA policies and procedures can be overwhelming and confusingly daunting. Find yourself a steady, reliable, and enthused employee to become your HIPAA officer in charge of compliance. It is very rare that employees wake up one day and magically become compliance specialists. Even those we put in charge of compliance have no idea what they have just got themselves into.

It is very easy to be the "Compliance Officer" on paper, but unless this person is trained or shown what the job entails, it becomes merely a paper title and compliance is doomed to fail. Often, the business owner will tell the office manager, "You're the HIPAA person," and hand them the HIPAA manual that's been sitting on the shelf for 10 years. They, in turn, will blow the dust off it, open it up and say, "You've got to be kidding; I don't have time for this." And so, the manual goes back on the shelf.

Partnering with a third-party compliance group provides training and assistance with enhancing the training experience while also encouraging a positive compliance culture that comes from knowing there is a third-party resource when questions arise. There are many organizations available to you that will assist in setting up your compliance.

The road to professional service and secure operation starts with compliance. It provides the safeguard for your office, your management, your employees, and your clients. An established culture of compliance that is embraced and enacted by the entire corporate hierarchy gives businesses the best chance to grow, enhance production, nurture a positive work environment, and ensure a positive customer experience.

Embrace it today.

About William

William Dickherber is the founder of Onsite Computer Consulting Services, which he started in 1995.

William is a Navy veteran with more than 40 years of computer technician experience. His penchant for fixing everything from tractors to televisions served him well when he was part of the pre-commissioning crew for the *USS Nassau*. He was instrumental in implementing the first computer-controlled communications system on a Navy ship.

He completed his associate degree in electrical engineering during his military training and went on to compete his bachelor's degree in Information Systems. When he left the military in the late 1980s, he took a job in the calibration lab for Emerson Electric. The computer software to which he was assigned did not work, so he wrote his own software to record the calibration status of company equipment. More than 25 years later, it was still being used there.

Not one to do things in a 'half-assed' way, William Dickherber started Onsite Computer Consulting Services when an organization wanted to hire him as a contractor rather than as an employee. He grew his one-man operation that operated out of a pickup truck into one of the region's most trusted businesses for managed IT services primarily for dental practices and small-to-medium businesses. He is working toward his virtual Chief Security Officer (vCSO) certification and is expanding his company's security presence in St. Louis.

Honesty and integrity are the watchwords for Onsite Computer Consulting. William pledges that he will not sell his clients anything they don't need. The company has earned local Business of the Year honors, and Dickherber prides himself on maintaining client relationships that he started nearly three decades ago.

William and his wife of 43 years, Ann, recently downsized to a 40-foot camper. They live with four dogs and a cat, all rescues. They have two children, a daughter who is the assistant director of housing for the University of Utah and a son who is a police sergeant in St. Louis and is currently working to complete his bachelor's degree.

The couple supports The BackStoppers®, which provides ongoing financial assistance and support to the spouses and dependent children of area police officers, firefighters and volunteer firefighters, and publicly-funded paramedics and EMTs who have lost their lives in the line of duty. William also supports CAPS, an immersive training program that immerses high school students in professional environments to help them choose a career path.

For more information, contact Onsite Computer Consulting Services:
- Email: office@computerparamedic.net
- Web: https://www.computerparamediconline.com
- Phone: 636-332-1335

CHAPTER 3

FOUR STRATEGIES NEEDED FOR CYBER SECURITY INSURANCE

FOUR STRATEGIES YOU MUST PUT IN PLACE BEFORE INSURANCE COMPANIES WILL EVEN CONSIDER WRITING A CYBER SECURITY POLICY

BY DANIEL WILLIAMS,
Founder & President – Xpedeus, Inc.

If you wanted a cyber security insurance policy for your business just a few years back, there was only one condition: money. Anyone willing to pay for cyber security coverage back then could certainly add it to their policy. But just as cyberthreats and ransomware have transformed the security landscape today, cyber security insurance policies have transformed over the years as well.

Cyber liability insurance policies first surfaced in the 1990s when they mainly covered online media and errors in data processing. In the 2000s, policies covered unauthorized access, network security, data loss and virus-related claims.

In 2003, California enacted the Security Breach and Information Act, requiring businesses to notify anyone whose personal information had been accessed by an unauthorized person. Across the US, other states passed similar laws. Suddenly, insurers began offering first-party coverage for IT forensics, information security and credit monitoring, as well as third-party coverage for regulatory defense, fines, and penalties.

AS CYBERCRIME SKYROCKETS, SO DOES THE NEED FOR INSURANCE COVERAGE

Today, the landscape of cybercrime has completely transformed, and the risks have increased exponentially from the 1990s and early 2000s. Consider these ominous statistics:

- Every 11 SECONDS, there is a new ransomware attack on an unsuspecting business. (Source: *Cybersecurity Ventures*)
- The cost of cybercrime is predicted to hit $10.5 TRILLION by 2025. (Source: *2022 Cybersecurity Almanac*)
- It takes an average of 287 days – nearly a FULL YEAR – for security teams to identify and contain a data breach. (Source: *IBM and Ponemon Institute*)
- A single cyber-attack costs companies an average of $200,000. And far too many of those companies go out of business within 6 MONTHS of the attack. (Source: *Hiscox Insurance*)

Because of the increase in cyber-attacks, malware, as well as ransomware today, insurance companies are being forced to develop aggressive new cyber-liability policies. These policies are designed to protect businesses from losses such as data destruction, extortion, theft, hacking, denial-of-service attacks, and ransomware.

However, insurance companies, by the very nature of their business model, must minimize risk. Therefore, they will only write cyber security policies for businesses that are adequately protected from data breaches. To ensure your computers and network are properly guarded from cyber-attacks, insurance companies require four cyber security strategies:

- Multifactor Authentication
- Password Management
- Phishing Training
- Cyber Security Training

WHY MULTIFACTOR AUTHENTICATION IS A MUST

At the top of insurance companies' requirements before they will write you a cyber security policy is multifactor authentication, also called two-factor authentication. Just as its name implies, multifactor authentication (MFA) requires a computer or network user to provide two or more verification factors to gain access to an application, online account, or a VPN.

Instead of granting instant access with just a username and password, multifactor authentication combines two or more independent credentials: what the user knows, such as a password; what the user has, such as a security token (physical device token, phone security app code and text message are just a few examples); and what the user is, by using biometric verification method. There is a small trade-off for the user, of whom slightly more effort is required, and that leads to a much more robust security model. It is far more difficult for a hacker to compromise a computer or network where MFA measures are installed.

At Xpedeus, our #1 goal is to protect your data and network. Therefore, all our clients will have our full security stack on their network or we're not for them. There's no way around it. Multifactor Authentication is one major component of our robust security stack.

Occasionally, a business owner or CEO will complain about the extra step necessary to log in to an application. I explain that those extra five or 15 seconds will make it considerably more difficult for a hacker to compromise your network and hijack your data. Besides, isn't an extra step better than days, weeks or even months of your business being dead in the water because you were a cybercrime victim? A slight sacrifice of convenience, it makes it 95% more difficult for cybercriminals.

The running joke is, "Do I also need an MFA code to open the door to use the bathroom?" While it hasn't quite got to that point yet, the reality is MFA will be increasingly required as the threat of network compromise continues to grow.

We believe our method for setting up MFA is far more secure. By putting all the multifactor authorizations inside a single security platform, there's just one place for all of the company's MFA happenings. If you release an employee, we simply disable that account and they no longer have access.

At Xpedeus, we will never concede when it comes to protecting your assets. Because if you do get ransomware or have your data stolen, while you may not sue us, your insurance company will. Therefore, we cannot make an exception. All our clients get our full cyber security stack, including multifactor authentication. Not only does that comply with our philosophy on security, but it also fulfills one of the major criteria insurance companies demand before writing a cyber security policy.

THE NEED FOR EFFECTIVE PASSWORD MANAGEMENT

A couple of decades into the new century, and what is one of today's most commonly used passwords? You guessed it! It's "PASSWORD!" Coming right behind that obvious answer, people still use "123456," "qwerty," their birthdate, their kid's name, or their favorite sports team. In other words, hackers only need to spend a couple of minutes learning about somebody to have carte-blanche access to all their data!

Just this year, TransUnion's South Africa unit was fully locked out of their own servers and ransomed for $15 million. How? Because their password was set as "password." A six-year-old could have gained access.

Just like with multifactor authentication, people today want speed and convenience when it comes to accessing their files, applications, and online content. To fulfill their ease of use, employees either have one single password for ALL their applications or keep all their different passwords in a spreadsheet. Believe me, cybercriminals know how to search your computer for these password gold mines.

At Xpedeus, everything we do for our clients, we also do ourselves. We have over 400 very important credentials, and each one has a separate password. They're all 32-character alphanumeric, including special characters and numbers.

To get a cyber security policy, here's what insurance companies may require from your password management:

- Long passwords. 15 characters or more.
- A mix of characters. Different letters, uppercase and lowercase, numbers, and symbols.
- Avoid memorable passwords. Anything connected to you (your pet's name, spouse's name, favorite sports team).
- Avoid sequential keyboard paths. Avoid "qwerty," "123456," etc.

So, how is one expected to memorize dozens of crazy-long passwords without a cheat sheet? You're not. In fact, we don't even know our own passwords. That's right, your employees shouldn't know your passwords either. Rather, you manage all of them through a trusted password manager application.

I tell my clients all the time that password management is not a technology issue. It's an administrative issue. We provide the electronic file cabinet to protect your assets. You simply have to file it and implement it correctly. So, when an employee is terminated, you make sure they no longer have access to your passwords. By adopting a sufficient password manager that's well managed and multifactor authenticated, you should have no problem getting a cyber security insurance policy.

YOUR FIRST LINE OF DEFENSE: EMPLOYEE PHISHING TRAINING

There's a reason many insurance companies today require comprehensive phishing training before writing a cyber security policy. According to Cisco's 2021 Cybersecurity Threat Trends report, phishing accounts for about 90% of all data breaches. And Verizon's 2021 Data Breach Investigations Report showed that 85% of breaches involved humans.

That means, like it or not, your employees are on the front lines of preventing a cyber-attack on your organization. Perhaps someone in sales accidentally clicks on an enticing e-mail. Or an administrative assistant receives an e-mail that looks like it's from their boss. Just one link, one click, and the bad actors are in. From there, the cybercriminals could hijack all your files, encrypt them, and demand a sizable ransom worth many Bitcoin. In fact, IBM's 2021 Cost of a Data Breach Report found that breaches resulting from phishing cost organizations an average of $4.65 million!

Gone are the days when you could easily spot a nefarious e-mail because of the broken English, weird symbols, and laughable subject line. Cybercriminals today are far more savvy. They can learn just how your boss writes an e-mail to mimic their voice. They may know you just ordered a MacBook Air and send an e-mail disguised as FedEx saying, "Click here to find the status of your package." Before your brain can even comprehend it, your finger clicks the link. That's when you immediately realize you just opened the front door for hackers to access your network.

We're human. We make mistakes. We get caught off-guard. That's why insurance companies are requiring MSPs and business owners provide phishing training before issuing cyber security policies. It's also why our team at Xpedeus provides comprehensive and ongoing phishing training.

We partner with a company that provides integrated security for training and phishing simulation. While we do provide in-person phishing training to business owners as well as their employees, the most effective training comes from simulations. By randomly sending out e-mails that resemble phishing scams but are perfectly safe, we quickly learn who in your organization is more likely to click on links. Those "click-happy" employees get more training and education.

With enough simulated phishing e-mails and ongoing training, your employees learn to question every e-mail. They become more guarded and more vigilant, resulting in much greater security to keep your network locked down and your data safe. That's just what insurance companies are looking for to keep from paying out crazy amounts of money.

PROTECT YOUR DATA WITH COMPREHENSIVE CYBER SECURITY TRAINING

Beyond phishing training, insurance companies will also require your business to have comprehensive cyber security training. An effective cyber security training program should consist of:

- Education about passwords, access privileges and secure network connections
- Device security for computers and mobile devices, especially if you have remote employees
- Cyber security threat response, which includes identifying a threat and containing it

At Xpedeus, we call it "cyber smart training." It's a holistic approach that provides an ongoing education that helps our customers protect their data. Our training answers the questions "What data do we have as a company that needs to be protected? What ways can our computers and network be infiltrated? How do we mitigate those risks by being smarter employees?"

A lot of cyber smart training boils down to simply thinking before you act. Just as you should consider who sent you an e-mail with a possible harmful link, you should also think twice before sending out a screenshot of a customer account number.

Our cyber smart training also encompasses compliance, such as the Health Insurance Portability and Accountability Act for medical and the Financial Industry Regulatory Authority regulations. Our medical clients go through a HIPAA compliancy training on an annual basis. When you sign on with Xpedeus and get our built-in cyber security, all your compliances should go pretty smoothly.

A WORD OF CAUTION

Anytime an insurance company can avoid paying out, they absolutely will. In writing the cyber security policy, the insurance company may not do their due diligence to make sure you have in place multifactor authentication, effective password management, phishing training, and cyber security training. However, they will absolutely audit your organization AFTER you've been hit. And just as in the case of a smoker who lied to get a life insurance policy, they will not pay out.

Don't worry, there's an easy solution. Rather than trying to achieve all these important cyber security measures yourself or running around to multiple vendors, Xpedeus handles it ALL for you. Not only will implementing these cyber security tactics make it a cinch to get covered for a cyber liability policy, but it will also go far to protect your data, your business and your finances from a cyber-attack or ransomware.

PROTECT YOUR DATA, BUSINESS AND FINANCES WITH A CYBER SECURITY INSURANCE POLICY

In our more than 21 years of providing IT services and cyber security to West Central Florida businesses, we have never had a client hit with a cyber-attack or ransomware (knock on wood). Even though our clients have been most fortunate, we still strongly advise getting a cyber security policy from a trusted insurance provider.

We learned about a local assisted-living facility that could not get cyber security insurance. Year after year, they failed to meet the insurance company's requirements. After signing on with Xpedeus and implementing our comprehensive security policies, they finally got the cyber security insurance. If your insurance provider refuses to issue a cyber security policy for your business, perhaps you should take a closer look at who is protecting your data. Maybe you are more susceptible to a cyber-attack than you thought.

About Daniel

Daniel Williams is the founder and president of Xpedeus, Inc., a two-decade proven IT, cyber security and telecom company serving West Central Florida. Dan received a bachelor's degree in Engineering and Business from Michigan State University and also pursued an MBA from Oakland University.

Dan has a rare acumen for technology, engineering, as well as business management, and a light bulb went on for him when one of his bosses said, "Williams, you will never be fully happy until you own your own business." Dan never forgot that statement. It lit a fire under him to gather as much technology and business experience as he could so he could one day fulfill that dream of owning his own company.

After college, Dan worked as a project manager for General Motors for nearly five years, where he gained valuable IT experience. After working for a multibillion-dollar company in GM, he became the project manager for a multi-hundred-million dollar company.

Still thinking about what his boss said to him, Dan knew it was finally time to start his own business. His current company had become less project-oriented and customer-focused and more about making a quick buck. That did not align with Dan's beliefs and his customer-first philosophy.

Unfortunately, 9/11 had just occurred. Through a cloud of fear and uncertainty, Dan Williams started Xpedeus, Inc., in December 2001 in Tampa, Florida. His former boss was right! Combining his love for technology and business was the life Dan had been searching for.

Dan's life fundamentally changed when he started reading the Bible and sought to learn what God wanted from him, becoming a Christ follower in his late 20s. His commitment to serving Jesus became his guide for how he would best serve his customers. As a result, for over two decades, Dan and his team have committed to always giving his clients a white-glove experience.

Many of Dan's customers have been with him for 10, 15 and even 20 years.

The reasons why begin with Xpedeus's ultra-responsiveness, including an average 14-second response time and an 18-minute average ticket resolution. It also speaks volumes that Dan's average employee tenure is over seven years.

After two decades of providing security, efficiency and affordability to West Central Florida, Dan Williams is poised to provide local businesses with the level of service they deserve for many years to come.

Get in touch with Dan:
- Email: Dan.Williams@xpedeus.com
- Phone: 813-430-4800

CHAPTER 4

UNDERGO HIPAA AUDITS WITH CONFIDENCE

BY CHARLES HAMMETT,
Founder, CEO & President, Hammett Technologies

I recently brought on a new client who works in behavioral health. She works closely with Special Needs Children. She reached out to me because her practice was growing, and she needed help to connect the staff between offices. Prior to speaking with me, she had thought she was Health Insurance Portability and Accountability Act compliant, after purchasing a $2,000 HIPAA online compliance questionnaire from a random person who cold-called her practice. The vendor didn't go through any of the 274 controls with her. They simply took her money, walked away, and left her with the impression that she was HIPAA compliant.

The idea behind the $2,000 online questionnaire is that someone at the doctor's office goes through it and answers all the questions. Then a report is generated that reveals whether or not you are HIPAA compliant. The problem with this approach is that when a doctor sees an acronym such as MFA or a question like "Do you require MFA on end-user devices?" they might not know that MFA means multifactor authentication – or that it could refer to a text to your phone, a random number on an app or a fingerprint scanner. Why would they? That's not part of a doctor's everyday life. A doctor has enough on their plate. They shouldn't have to customize policies and procedures, implement the controls, and fill out HIPAA compliance forms. What this company sold my client was essentially an online checklist.

I'll never forget the look on my new client's face when I told her she wasn't even close to being HIPAA compliant. Besides the terrible feeling of being duped out of her money, she also had to come to grips with the fact that her business and patient data were not properly protected. I wish I could say that what happened to this woman was exceptionally rare, but it's not. There is more than one company out there selling these online HIPAA "checklists" to businesses and leaving them with the impression that handing over $2,000 is all they must do to become HIPAA compliant. These companies send out flyers to medical offices that talk about the importance of being HIPAA compliant ASAP and urge the reader to make a phone call. Doctors know they need to do something, so they make the call, and the misleading sales pitch begins.

A RECIPE FOR FAILURE

But the problem isn't just deceptive sales pitches – some MSPs are falling short of their HIPAA responsibilities. We recently brought a chiropractor onboard as a client. Their former vendor had told them they were HIPAA compliant. The first scan we ran revealed that they missed many required controls. Even some of the simple ones, like patching the machines, had not occurred in months. Their vendor had told them that everything was covered. However, they never provided an assessment or a report that showed the client what exactly they'd done for them in that regard. There was no documentation. No accountability. Nothing. Often, the person in charge of compliance at the covered entity (CE) is a family member or a patient who works in an IT department and is helping them. Neither specializes in health care compliance nor understands how to put the controls in place properly. Having the wrong level of expertise in the right place is a recipe for failure.

In both situations, I'm convinced that if someone from the Office for Civil Rights (OCR) phoned them up and told them they were going to be audited for HIPAA compliance, they'd say they were confident they would pass. That said, HIPAA compliance audits are not regularly performed on all businesses operating within the health care field. While it might be a good practice financially, the OCR can't go out and audit all the CEs and business associates to ensure that they're adhering to HIPAA standards.

IGNORING HIPAA COULD DEVASTATE YOUR BUSINESS

But audits do happen. And when you fail, or you have a data breach, the results can devastate your business. The minimum fine for a willful criminal violation is $50,000, according to the *HIPAA Journal*. The maximum criminal penalty for an individual is $250,000. You might also have to pay restitution to the victims, and you could even face jail time. Civil penalties start at $100 per violation, which could rise to $25,000 if there are multiple violations of the same type. (Many variables go into determining how much the fine will be, such as the nature of the violation, whether it occurred because of neglect, what corrective action was taken, the harm caused and the number of people affected, whether rules were violated for personal gain and so on.)

HIPAA compliance is nonnegotiable. It is your responsibility to ensure your practice is meeting HIPAA regulations so that if an audit or an assessment happens, you are confident walking into that meeting knowing you have the documents, processes, and policies in place to meet or exceed their requirements.

Here are the five primary reasons you might receive a call informing you that you are going to be audited:

- The OCR receives a complaint against your organization. This is the most common. It could be a patient in your building who sees something you're doing with their protected health information (PHI) that they don't like. It could be a competitor or a friend of a competitor who complains about something to put undue strain on your business. The complaint could also come from a disgruntled employee or ex-employee.
- Your business is the target of a data breach, which (as per OCR guidelines) you reported to the OCR.
- A partner, vendor or business associate is breached, and you become part of the investigation by the OCR.
- You're randomly selected to be audited by the OCR.
- If you are a new entity in the health care space, they *might* require you to go through an audit before they allow you to treat patients. This happened to one of my clients recently.

Alexander Graham Bell once said, "Before anything else, preparation is the key to success." The key to undergoing a HIPAA audit with confidence *is to prepare*. To help make sure you're prepared and ready, I've put together five steps that will get you where you need to be in terms of HIPAA compliance:

1. **Choose a security and privacy officer.** Someone in your organization must be responsible for demonstrating the steps that are needed to protect and keep private all the PHI in your organization. This person will schedule periodic reviews and do a risk analysis. They should be able to decipher the data and the logs. Your security and privacy officer records all data breaches and incidents and, if they meet a certain level, reports them to the OCR. This security and privacy officer will be in charge of reviewing all agreements with your business associates, your product vendors, and your Internet service providers to make sure they meet HIPAA compliance. If you're hosting your electronic health records

on a HIPAA compliant hosting service, the security and privacy officer needs to assess them to verify that they are HIPAA compliant. Your security and privacy officer should also validate that your fax and your email service meet HIPAA compliance standards and that every location where you're storing health care data is meeting the HIPAA compliance controls and guidelines.

The ideal person to assign the title of security and privacy officer is someone who has a good overall knowledge of the area of health care they are working in and is familiar with the way the practice operates. The security side of it requires they possess a strong technical background because they're going to be the individual who is testing the security of the electronic protected health information (ePHI) controls. They need to understand how to read, decipher and implement the controls. I cannot stress enough the importance of choosing the right person to be your security and privacy officer. It's a key part of preparing for an assessment. They are going to be involved in every audit meeting and they will answer most of the questions.

2. **Analyzing and reviewing your current policies.** The HIPAA guidelines clearly state that everything within your practice needs to be thoroughly documented, and documents need to be accessible for everyday use within your business. Your policies and procedures must align with your technology. Questions such as "How do we patch the machines?" "How do we back them up?" "How do we secure the data?" must be answered. If they aren't answered or they don't align with the HIPAA guidelines and controls, you must rewrite your policies. Getting your policies right takes time. Some can be taken care of quickly, while others could take months to complete.

If your business doesn't have a technology plan, this is the perfect time to create one. A technology plan is a detailed description of not only the goals of the organization around technology but also shortcomings and issues that need to be reviewed, addressed, and resolved by the leadership. Everything needs to be documented with a detailed description of each issue, the start date, the stop date and who is responsible for it. It's critical that there is accountability every step of the way.

3. **Perform an internal audit.** The HIPAA guidelines suggest you do an annual internal audit to make sure you are compliant. This is an audit not done by OCR or a third party, but one you do on yourself. This internal audit will show you where you are with HIPAA compliance, and it will pinpoint any areas where you are falling short of HIPAA's compliance requirements. All the audit findings should be added to your technology plan. This way, you can use one document to drive the meeting to help mediate any issues that arise. Findings need to be reviewed with all stakeholders and leadership. These findings will determine whether additional funding, products, people, security, training, or implementation timelines are required. After your audit, schedule your next audit.

At a minimum, bring in a third-party company every three years to do an audit. It's important to get an honest, unbiased snapshot of where you are with your security and compliance. The third-party company will look at your infrastructure, your storage, your backups, your workstations, your employee workflows and where your ePHI is stored – and they will validate that everything is meeting HIPAA compliance standards. This audit will reveal any shortcomings you might have and provide you with a list of action items to get you to where you need to be.

Note: We do these annually for every health care client whose systems we manage. We run an audit tool monthly to collect the data we use for the internal audit reports we present during our business reviews. So, if you partner with an IT company such as ours, you don't have to perform an internal audit annually because we analyze and monitor your situation on an ongoing basis.

4. **Remediating the audit findings**. All your internal audit findings should be reviewed with the company stakeholders, whereupon you should develop plans to remediate each issue in a timely fashion. Some factors will drive the remediation timelines, such as the necessary training, the resources available and any required investment that will be unique to each practice. All issues and changes must be documented. This saves everyone a lot of time and makes it easy to determine what approvals (budget or otherwise) are necessary to move forward. A recurring meeting should be set up with the stakeholders and the technical lead to ensure that everyone is doing their part to meet the deadline they are accountable for and that they are adhering to the technical plan. Meeting notes should document everything discussed.

5. **Educate your employees**. HIPAA guidelines require all employees involved with the processing of PHI or ePHI to be adequately trained to maintain industry-standard privacy and security measures. When changes occur to ePHI laws, your employees must be trained on these changes. They should be trained if changes are made to the system or to how security is handled within the organization. Records need to be kept, proving to the auditors that the training has been performed. The easiest and most efficient way to keep your employees trained is to create standard operating procedures and guidelines. The SOPs and guidelines need to be updated on an ongoing basis. The security and privacy officer should be given the responsibility of keeping track of all the information and ensuring your employees always have the most up-to-date information. In addition, it's important to convey to your employees why HIPAA compliance is so

critical and how adhering to these controls and practices will boost the health of the business and keep them gainfully employed.

You must prove to the auditors that your employees have undergone HIPAA training. So it's important to keep track of who has been through what training. If someone cannot answer the questions at the end of the training, make sure they go back and do the training over until they're able to answer all the questions correctly. A report showing who has completed the training should be supplied to the leadership of the business.

Initially when we educated people on HIPAA controls and procedures, we would have everyone sit in a room for an hour and a half to go through the various topics and issues. But we found it was too much information all at once. People had a hard time focusing, and they didn't retain much of the information. When we do the training weekly for about 10 to 15 minutes at a time (usually with an informational video), it's easier for people to digest and retain the information, and people never find it boring.

Annually, I host a security presentation with a local chamber that outlines what a company should do to keep its business and client data safe from cyber and ransomware attacks. At the close of my discussion, I'd look out into the room, and everyone would be in shock and sweating. Although I didn't mean to, I scared many of them. Plus, many were suffering from information overload. Internet security and HIPAA compliance do not have to be scary. In fact, it can be downright easy if you partner with an MSP who specializes in HIPAA compliance. However, if you hire the wrong business, the process could be burdensome. Unfortunately, there are a lot of companies out there that don't understand the complexities of compliance. Some companies will tell you what to do, but they don't know how to do it themselves.

To save money and achieve all your business goals successfully, you must partner with a company that can provide you with the entire solution. Someone who can assess your situation, help you remedy any deficiencies and then report on every aspect of what transpired.

CHOOSING THE RIGHT PARTNER

So, how do you evaluate a business to help you become HIPAA compliant? The first thing I would do is have them run through their process of how they will assess you. What are the key issues they will look at? They should supply you with a pre-assessment questionnaire. It should contain questions such as:

- "Do you have a technology plan?"
- "Where do you store your ePHI?"
- "How do you handle the electronic medical records?"
- "Do you have a system security plan?"
- "Have you completed a risk assessment? If so, when was it completed?"
- "How do you monitor your systems?"
- "How do people access company resources internally and remotely?"
- "Do you have baseline configurations?"

These are just a few of the questions we ask during our discovery calls. Insist they go step-by-step through their process (it should be similar to the five-step process I've laid out here) and have them explain each step and why they are doing it. How they answer these questions and explain their process will allow you to gauge whether they have the knowledge necessary for you to trust them with this important aspect of your business.

Also, this is of paramount importance: make sure you have a good feeling about the type of relationship you will have with them. Don't just hire them on the spot. Have a meeting with them. They are going to know a lot about your practice, so it's important you not only like them, but you trust them. When you're in evaluation mode, it's a good idea to have multiple meetings with multiple vendors. Interview them as you would an employee to make sure they are a good fit.

PREPARATION EQUALS CONFIDENCE

Around the 2017 time frame, when I would speak to a business about HIPAA compliance, their attitude was nonchalant. Many were aware of HIPAA but felt that if they didn't know the details of it, they wouldn't have to do anything about it (the old – and often faulty – "What you don't know can't hurt you" scenario). Over the last five years, though, attitudes have changed and matured. More and more CEs are aware that they need to be compliant. They realize the negative effect ignoring HIPAA could have on both their bottom line and their reputation. With the White House putting out an executive order on cyber security, awareness and understanding of the importance of HIPAA compliance are only going to increase.

In conclusion, if you do the preparation, you do not have to fear a HIPAA compliance audit. If you follow my five-step process, you will go through a HIPAA compliance audit with confidence and peace of mind that you will pass with flying colors. I highly recommend that to make the process go smoother, you leverage a third-party company that specializes in HIPAA compliance to help you through the process.

About Charles

Charles Hammett is the founder, CEO, and president of Hammett Technologies, which serves Maryland, Delaware, and Virginia. He is a Microsoft Charter member who has multiple high-level industry certifications (MCSE, MCSA CCEA). Charles has over 25 years of experience dealing with Health and Human Services (HHS) and the Centers for Medicare and Medicaid Services (CMS) and adhering to their best practice system security model. In addition, he has over 15 years of experience in implementing NIST security controls and the policies behind them. Hammett Technologies specializes in HIPAA compliance, with Charles developing a set of tools and scripts that help automate the assessment process, which saves his clients time and money.

Charles founded Hammett Technologies (under the name MDPCHelp.com) in 2002 as a side business while he was working for CSC, an HHS and CMS government contractor. He started his business because a friend who owned an awning manufacturing business (and is still a client today) needed some technical expertise. Working full-time during the day, Charles discovered his clients preferred him working on their IT at night because it didn't interrupt their daily business activities.

From the start, the driving force behind his business has been to provide smaller businesses he felt were underserved and overcharged by many of his competitors for a high level of service (on par with the level of service he provides larger clients). He witnessed firsthand how some of his competitors were forcing products and capacity on their clients that weren't needed. He believed they focused too much on how much money they could make, instead of providing their clients with the best possible solution.

Through word of mouth and networking, after six years and still working part-time at night, Charles grew his IT business substantially. To make sure his clients were well-served, he brought on his son, Garrett, to help him manage his clients as he continued working full-time for another government contracting company. In 2012, he focused solely on MDPCHelp. com, which he rebranded as Hammett Technologies, LLC, a fully-outsourced IT management company. He also streamlined their product line to increase their level of expertise and reduce their clients' support footprint. Shortly after the rebranding, Charles won substantial government contracts from the HHS and the Transportation Security Administration.

Today, Hammett Technologies has two fully operating IT teams: one is dedicated to health care and the other to manufacturing in the construction arena. They also have a team that runs a Baltimore data center for HHS. Two things that have proved extremely popular with his clients are his "all in one" pricing (where clients pay only for the services they are using) and that each client has one point of contact for their account. Charles's goal is to make technology and HIPAA compliance the least complex area of his clients' businesses.

For more information, contact Charles at Hammett Technologies:
- Email: charles.hammett@hammett-tech.com
- Phone: 443-216-9999; 1-877-659-4399
- Web: https://www.Hammett-Tech.com

CHAPTER 5

THE MOST COMMON HIPAA VIOLATIONS
THAT EVERY HEALTH CARE BUSINESS SHOULD KNOW ABOUT

BY CHRISTERBELL (CHRIS) CLINCY,
President, Simplified Cyber

The Health Insurance Portability and Accountability Act of 1996 (HIPAA) ensures the privacy and security of patients' electronic protected health information (ePHI). This law not only enforces health status confidentiality but also prevents criminals from accessing personal health information traversing the Internet, on mobile devices or via video-conferencing platforms for remote medical visits.

Doctors, clinics, dentists, psychologists, nursing homes, hospitals, insurers, and other health care providers, as well as their business associates, like billing companies, accountants, law offices and online patient portals, are required to follow HIPAA requirements.

However, not everyone adheres to the rules.

The Department of Health and Human Services' Office for Civil

Rights has settled with violators to the tune of millions each year. Settlements involve more than money. The government can also inspect your business and force changes to make it compliant for years.

How do you avoid invasive governance, paying painful financial settlements and suffering the shame of notifying patients of a breach? It starts with knowing common HIPAA violations. Does your internal IT team or Managed Services Provider (MSP) know how to protect against these violations so ePHI can stay secure, especially after rules were relaxed for the Covid-19 pandemic?

TELEMEDICINE AND THE PANDEMIC

At the onset of the Covid-19 pandemic, the health care industry and the government came together to make sure health care continued to be available to everyone, especially high-risk individuals. HIPAA loosened regulations to allow the rapid onboarding of vendor-supplied telehealth solutions and Medicare, some privately owned insurance companies extended coverage to remote patients, and mobile health applications hit the markets at breakneck speeds.

While telehealth made it possible for people to get medical care during the Covid-19 pandemic, it also brought to light three shortcomings that show health care companies are an easy target for cybercriminals, and why these companies need to make an investment in their IT security:

1. Many health care companies and practices do not have the IT and cyber security chops to adequately implement a secure IT infrastructure to keep ePHI safe.
2. Telehealth implementations have created new points of entry for attackers, all of which need to be protected in order to keep personal health information secure.
3. When the health care provider may have had good cyber security hygiene, their less-savvy patients did not.

As a result, telehealth represents a huge cyber security risk for health care for both the provider and the patient. In fact, SecurityScorecard named telehealth the biggest health care threat. Their research found evidence of threat actors selling ePHI data, malware toolkits and ransomware uniquely configured to take down health care IT infrastructures.

There are even malware applications hidden to look like legitimate telehealth apps that have caught companies unaware. And there are the usual web app vulnerabilities or the security holes opened through poorly-configured applications.

Knowing how to identify these telehealth security concerns can be tricky, but your local MSP can help. Investing in an MSP or in-house cyber security professionals to safely implement these new and emerging technologies will boost the cyber security of health organizations, improve HIPAA compliance, and inspire the trust and confidence of patients.

WHY HACKERS ARE AFTER ePHI

It's well-known in the cyber security community that data from medical-related businesses gets much more money on the dark web than credit card information. Do you know why?

Credit card theft is easy to recognize by the victim. It shows up on a statement or gets discovered during a transaction. Once the victim notifies the bank, it's simple to disable that card's information from further abuse. But hackers can abuse ePHI for years before victims find out. Therefore, hackers who want to commit identity theft will choose medical companies over credit card information if they can.

Electronic PHI includes a patient's full name, date of birth, address, phone number, email address, social security number, insurance ID number and full facial photos. With so much personal information, a nefarious actor could commit medical identity theft to get loans, file for retirement and even receive free medical care.

Medical identity theft can also cause great harm to the victim if false entries in their health records lead to incorrect medical treatment. Likewise, medical providers could get in trouble with the law if their provider number is used to bill for treatment or narcotics they did not authorize.

HOW ARE HACKERS GETTING ePHI?

Malware or ransomware attacks are common methods hackers use to infect hospitals' and medical practices' computer networks and gain access to personal medical information, and these attacks are not going away. They will become even more sophisticated as doctors, clinics and hospitals continue to embrace technologies like telemedicine to help manage their patient's health.

In fact, the FBI's Cybersecurity and Infrastructure Security Agency (CISA), along with the Department of Treasury, released a Cybersecurity Advisory on July 6, 2022, about a ransomware system called Maui that's targeting the health care and public health sector.

These government agencies discourage health care providers from paying ransom payments, but many do because regaining access to that information can save the lives of patients. Research by Sophos, a major cyber security company, revealed that "on average 61% of health care organizations pay the ransom, and of those organizations that paid the ransom only 2% got the data back." Thus, cyber security at your business goes hand-in-hand with HIPAA compliance. Good security practices can save your data and it can save the lives of patients.

Now that you know how serious the problem can be and how unwilling cybercriminals are to give your data back even if you pay a ransom, let's look at the most common HIPAA violations.

WHAT ARE THE MOST COMMON HIPAA VIOLATIONS?

HIPAA is a complicated law, but the part that is most important for cyber security is the Security Rule. The Security Rule came into effect on April 21, 2003, as a response to professional use of personal mobile devices within the health care community.

The Security Rule of HIPAA states that covered entities must "implement administrative, technical, and physical security controls" to protect ePHI and its exchange. The most vulnerable of the three are the technical controls. Many of the violations in this chapter fall into that category.

There is no doubt that going digital has many benefits for accessing and protecting health information, but if these controls aren't there, your organization is at risk of hacking attacks and HIPAA violations.

When we consider all the technological advancements in health care since 2003, it is easy to appreciate the guidance HHS provides in setting standards that prevent unauthorized disclosure of ePHI. Here are the most common violations of the Security Rule:

Breach Notification and Risk Assessment

When a breach happens, companies must perform two tasks:

- Perform a risk assessment to determine how the breach happened, why it happened, how bad it was, and how to fix it in the future.
- Notify the affected patients within 60 days of the breach so they can take steps to protect themselves.

These are requirements under the HIPAA Security Rule of 2003 and the Health Information Technology for Economic and Clinical Health Act of 2009. The purpose of the risk assessment is to address the likelihood of any further risk from becoming realized by removing vulnerabilities and blocking cyberthreats from jeopardizing ePHI. We like to refer to this as proactive maintenance.

Premera Blue Cross had to pay a $6.85 million settlement for risk management and risk analysis failures. Excellus Health Plan had to pay $5.1 million. Doctors in private practice must also comply and could get fined $100,000 or more. This is serious business!

Also, the Strengthening American Cybersecurity Act of 2022 requires medical providers to report breaches to CISA. By reporting breaches to the federal government, you can help America improve its cyber security and receive help to improve your network. To make a report, visit https://cisa.gov/report.

Unprotected Health Information

Once ePHI is on your network, you must protect it. Creating policies and enforcing those policies using technical and physical controls, and training employees on how to recognize phishing attempts, will help keep ePHI safe.

We have found that most medical providers do not have the time to keep current with the latest cyberthreats and their defenses, including employee education. An MSP's assistance is invaluable in helping defend against today's advanced cyber-attacks. Some strategies you should expect from your MSP are:

- Limiting access to data by deploying encryption and digital certificates
- Prohibiting the use of admin accounts for day-to-day activities and leveraging multifactor authentication
- Turning off Internet access when it's unnecessary
- Employing multilayered network segmentation for specific data types
- Encrypting data at rest and in transit using strong protocols
- Protecting patient data on internal systems behind a firewall

- Deploying monitoring tools to observe whether medical devices are behaving erratically due to a compromise
- Creating and regularly reviewing internal policies
- Protecting PII and PHI, as outlined in HIPAA

As long as ePHI is valuable, there will always be thieves trying to steal it. Creating a strong security policy and enforcing it will go a long way toward proving due diligence and due care in the event of a breach. If your medical organization cannot prove due diligence and due care, the government can hold you liable for negligence.

Once these controls are in place, don't budge on them. The small value achieved by loosening security in favor of productivity is far outweighed by the fines and embarrassment of a breach.

Failure to Use Encryption

The Omnibus Final Rule of 2013 added the use of encryption to render ePHI undecipherable in the event of a breach. However, HIPAA encryption requirements are vague, stating that companies must "implement a mechanism to encrypt PHI whenever deemed appropriate." Most engineers interpret this to mean that if the data is being transmitted over the Internet, it must be encrypted.

However, if data is stored on your internal network with no access to the Internet, then based on that risk analysis, you may determine that encryption is unnecessary. Many companies use this interpretation to decide that encryption is optional. It's not. There is no downside to using encryption for your data.

What happens when you don't encrypt? Lifespan Health System Affiliated Covered Entity had to face a penalty of over $1 million for an encryption violation that exposed ePHI of over 20,000 patients. When that much is on the line, the costs of implementing and enforcing encryption on drives are negligible.

Once you encrypt ePHI, this data is of no use to any third party that doesn't have authorization. It is the best way to achieve transmission security. But that doesn't mean it's easy to implement in the beginning. From our experience, we have found that medical providers' environments that are reliant on BYOD policies have the most challenge in maintaining encryption. Your MSP or local IT professionals can provide mobile device solutions that can secure corporate data on personal devices.

Unencrypted USB devices holding ePHI are also problematic, but they can be managed. Microsoft Windows and MacOS operating systems offer several options for managing encryption on USB devices. For example, USB drives can be configured to encrypt USB devices once inserted. You can also disable USB ports on your internal systems so the drives can't work.

The Mann-Grandstaff VA Medical Center lost two USB drives containing ePHI and exposed the information of almost 2,000 veterans. These drives were storing data from a decommissioned server and weren't wiped. Seven months later, they were stolen during a service call. The veterans had to be notified and given credit monitoring services for a year.

Later, the same company lost an unencrypted laptop computer used in a hematology analyzer. That breach exposed the names, dates, and SSNs of 3,200 veterans. The company created a system that allowed remote sanitation of devices after that.

Data and devices are just the beginning. You could also create digital certificates to authenticate connections to an electronic health record system, while storing the secret key in a FIPS 140-2 certified hardware security module and require the use of a VPN for any remote work.

Encryption may sound like a technical topic, but it's something that IT professionals have studied for decades. Any competent MSP can help you get your systems properly encrypted. It's an extremely strong technical control that anyone storing ePHI should employ to protect patient information.

Improper Disposal of Patient Data

Once PHI and ePHI are no longer needed, HIPAA requires that the information be destroyed. Destruction of storage devices through shredding or using secure wiping software will prevent electronic data from getting leaked to the public.

Most MSPs can handle data and storage destruction for you. If you plan to resell your old IT equipment, you will need a secure wiping solution and a way to confirm it worked before you sell it, to prevent ePHI exposure.

Most of the public exposures of nonelectronic PHI involving improper disposal involve paper records. Leaving unshredded documents in a trash bin is enough to trigger a violation. If the information can end up in a place where the public – even a dumpster diver – can reach it, that's enough.

We hear stories all the time about mobile devices containing ePHI being traded. If it's found and reported, it can trigger a violation and lead to fines. Make sure you're wiping or shredding your old storage mediums!

Employee Misconduct

Employee misconduct around medical records and private health information can also expose a business to liability. For example, it's tempting for a worker to access a record about a celebrity that stayed at your facility. Without a good reason to do it, this violates HIPAA.

Social media misconduct is also a growing HIPAA problem, and one that's not as easy to defend against. Here are some examples:

- A group of resident doctors in the Spectrum Health system in Grand Rapids took photos of body parts removed from patients, including on the operating table. While names weren't revealed, there could be enough information to cause a HIPAA Privacy Rule violation.
- A Facebook group with over 23,000 EMS professionals shared private patient health information while mocking patients. Later investigation found that Facebook had shut down that group twice, but it had reformed.
- A dental office disclosed PHI on Yelp in response to a review. It exposed the patient's name, treatment plan, insurance, and the cost of the procedure. A government investigation found another similar response. The government fined the office $10,000.
- A Snapchat post showed two employees at a Glenview nursing home taunting a 91-year-old with dementia.
- A hospital employee at UC Davis Medical Center posted an update on a toddler's condition on a news story on Facebook after her passing. Later, an internal investigation found that the worker accessed the data improperly.

Employers cannot control every aspect of their employees' lives, but they should have strong internal procedures for handling breaches like these, including how to inform patients who had their information exposed and how to discipline employees who infringe like this.

THE BEST WAY TO PROTECT YOURSELF

How do you protect yourself from these common HIPAA violations?

Health care organizations must comply with HIPAA rules more strongly now than ever before. Internet-connected technologies have allowed health care providers to improve health care, but this introduces new cyberthreats, resulting in cyber-attacks that cannot be ignored. It starts from the top down. Companies must implement administrative, technical, and physical security controls to protect ePHI and its exchange.

Sometimes the bad guys will get lucky despite your best efforts. However, a risk assessment, appropriate cyber security defenses and timely notifications will go a long way toward proving due diligence and due care in the event of a breach.

Anyone handling PHI – even small practices – needs to comply with HIPAA, but what if you lack a compliance or IT background? How can you effectively protect yourself and your business? The answer is to use an MSP to handle it for you.

The job of an MSP is to manage IT problems on behalf of companies. This includes creating policy, educating employees about cyber security and HIPAA risks, providing cloud infrastructure that meets HIPAA standards, encryption, access controls, audit controls, physical safeguards, transmission security and much more.

Most of our clients do not have dedicated staff to commit to HIPAA compliance and opt to hire us to do that work for them. Hence, I recommend working with a HIPAA compliant security-focused MSP. This is the simplest way to comply with HIPAA obligations.

HIPAA-focused MSPs know the rules, how to report compliance and what needs to happen in the event of a breach. They can provide solutions that include everything you need to satisfy the Security Rule and other HIPAA regulations. If a violation is discovered, they will also be on the hook to explain what happened and could become the target of any negligence claims.

About Chris

Christerbell Clincy is the owner and president of Simplified Cyber, a Managed Services Provider that helps small to medium-size businesses simplify their security requirements by leveraging existing technology. She currently works with companies and health care providers to ensure they are meeting both security and compliance needs.

She is driven by the desire to make the best security options available to everyone – not just big companies. This is increasingly necessary as cybercriminals are getting smarter and using more sophisticated, cloud-based computer systems. It is not an even fight compared to the on-premises technology used by most businesses.

Chris says technology is evolving to include more simple, affordable security options, which is the direction Simplified Cyber is taking their customers. One of their priorities is providing tools that secure the network perimeter for those who work from home, or places like coffee shops, where they are connected to public WiFi. This can be resolved with technologies that offer a small bit of software that can be loaded on a system to give protection from any location.

Chris embraces the 'Cloud-First' world and even shares her knowledge with others in parts of the world with fewer resources. She volunteers with well-known groups that provide free IT and cloud training to developing countries like a few on the African continent.

She also served in the US Army, in data communications, deploying tactical networks before units arrived for training. She and her team set up routing, switching and email messaging services in the field. They prepared for war, training as if they were in a hostile situation, making sure the underground soldiers could communicate with the Air Force in the sky. Christerbell's role was to design and deploy computer network infrastructures to include routers, switches, firewalls and IDS/IPS systems within the perimeter; then, when training was over, she pulled all the cabling and infrastructure.

She still uses some of the security measures she practiced in the military, such as making sure hard drives are encrypted and firewalls are in place and blocking unwanted traffic with access controls appropriate to individual roles and clearance. This experience fortified her ability to see several steps ahead while working on any difficult problem.

Chris has worked in IT security and systems management for more than 25 years. She is proud to have worked in nearly every role, including technician, systems administrator, engineer, cloud solutions architect, project manager and instructor. She is a technology evangelist and loves having a breadth of knowledge to share with individuals and organizations of any size. She feels lucky to have a career that is also her favorite hobby and to have a husband who supports her.

For more information, contact Chris at Simplified Cyber
- Email: info@simplifyyourcyber.com
- Phone: (800) 931-5785

CHAPTER 6

TAKE THE CONFUSION OUT OF HIPAA COMPLIANCE

BY CRAIG PETERSON,
Founder, Digital DataComm

The Health Insurance Portability and Accountability Act of 1996 (HIPAA) went viral in the news and on social media during the pandemic. Many public figures cited it as the reason for not answering questions about whether or not they had received the vaccination for Covid-19, stating that to do so would be a violation of HIPAA. To paraphrase one of the most quoted lines in the classic film *The Princess Bride*, people keep using the word, but it doesn't mean what they think it means.

At the root of confusion regarding HIPAA is grappling with understanding what HIPAA is, how it applies to entities, and what needs to be done to be in compliance. This lack of understanding, I have found, comes from the attitude "I don't feel like we have a liability – we are not at risk."

As anyone who has read the HIPAA manual can attest, it is not exactly a page-turner. The dense legalese can create confusion around grasping its basic concepts. But it is imperative that everyone in your business know their role in HIPAA compliance. The consequences for a lack of HIPAA knowledge and training may include data breaches that imperil patient privacy and your company's reputation, prohibitive financial penalties, employee termination and, in the worst cases, jail time, bankruptcy and business closure.

Even after you attain an understanding of HIPAA's importance, there is often confusion over what protections should be put in place, how data should be properly stored and other protections to mitigate risk. Many times, a business owner will rely on a vendor without checking to see if that vendor is really following HIPAA compliance when it comes to software programs they are using to store the patient data.

Many people wrongly define HIPAA as a generic term for health privacy. But HIPAA is a federal law that created national standards to protect sensitive patient health information from being disclosed without the patient's consent or knowledge, according to the Center for Disease Control. Under HIPAA, all patient and employee information, including that pertaining to past or present medical conditions, treatments or therapies received and payments made for medical care or treatments, are considered private and confidential.

To implement the requirements of HIPAA, the U.S. Department of Health and Human Services (HHS) was charged with issuing the HIPAA Privacy Rule in regard to protected health information (PHI) while allowing the flow of health information needed to provide high-quality health care and to protect the public's health and well-being. The HIPAA Security Rule establishes national standards to safeguard an individual's protected health information both written and electronic (e-PHI) that is created, received, used, or maintained by a covered entity. The Security Rule requires several safeguards to ensure the confidentiality, integrity, and security of written and electronic PHI.

HIPAA covers four primary entities:

1. **Health care plans**, which include health, dental, vision and prescription drug coverage. (Employers that establish, maintain, and administer group health care plans for fewer than 50 participants are excluded from this regulation and, according to the government, are not considered a covered entity.)
2. **Health care providers**, who transmit any health information in electronic form in connection with transactions such as claims, benefit eligibility inquiries, referral authorization requests and other transactions that fall within the HIPAA Transaction Rule.
3. **Business associates**, which are individuals or organizations not already part of the covered entity that use or disclose individually identifiable health information in performing or providing services or activities on behalf of a covered entity.
4. **Health care clearinghouses** that process services to a health care provider or a health plan.

HIPAA does not apply to schools and businesses outside the context of health care, but even if they do not have a direct connection to the health care industry, a company that offers health insurance, a Flexible Spending Account plan or a wellness program among its employee benefits must comply with HIPAA security and privacy requirements. (Unless, as stated above, the company has fewer than 50 participants.)

In getting a handle on HIPAA, there are several key points on which to focus. The first concerns PHI. This is the data a health care professional collects to identify an individual and determine appropriate care. It includes the patient's name, address, contact information, Social Security number, photographs, and medical record and health plan beneficiary numbers. It also includes medical histories, test and laboratory results, mental health conditions and insurance information. Businesses need to enact a plan that ensures there are safeguards in place to protect this information, regardless of how it is stored.

As HHS.gov notes, the health care industry did not have a generally accepted set of security standards or general requirements for protecting health information prior to HIPAA. Meanwhile, the industry was moving away from paper processes and relying more on the use of evolving electronic information systems for such tasks as claims payments, answering eligibility questions, providing health information, and conducting other administrative and clinically-based functions.

There are three safeguards mandated by the Security Rule for the security and privacy of PHI: administrative, technical, and physical.

The Security Rule defines administrative safeguards as "administrative actions, and policies and procedures, to manage the selection, development, implementation, and maintenance of security measures to protect electronic protected health information and to manage the conduct of the covered entity's workforce in relation to the protection of that information."

Physical safeguards pertain to policies and procedures enacted to protect against inappropriate and unauthorized access to protected data by controlling access to facilities, rooms, equipment, and desks where PHI information might be present or stored, including protection from natural and environmental hazards.

Technical safeguards refer to how your business restricts access to computer systems and protects them and the communications containing PHI transmitted electronically over open networks from being intercepted by anyone other than the intended recipient.

Background and credit checks should be performed on those individuals who have access to PHI and reviewed by human resources and management.

HIPAA and cyber security go hand in hand, and another common question involving HIPAA compliance is what data protections a company should enact. I am often asked how much companies spend on average on HIPAA compliance and cyber security. My answer is always, "Not enough."

It is essential that companies enact cyber security training for their employees, who, unwittingly, often prove to be a company's biggest risk to data protection. For example, if an employee clicks on a phishing email, they may have inadvertently given computer or email access to proprietary data.

Employee cyber security training should focus on what HIPAA is and how to protect your company's data from cyber-attacks, such as the above-mentioned phishing (phishing simulations can be invaluable training exercises).

HIPAA requires that covered entities conduct a risk assessment of their health care organization to ensure it is compliant with administrative, physical, and technical safeguards, and to reveal areas in which PHI could be at risk.

But whether or not that is required under HIPAA, cyber security measures come down to common-sense best practices. For example:

- Are antivirus and anti-malware software running on all your computers and servers, and are they centrally monitored?
- Where possible, have USB drives been disabled from computers? This will prevent bad or unwitting actors from extracting proprietary data.
- Are employees using email encryption when they correspond with someone with health care information, such as an invoice with classified HIPAA data?
- Are patients' records encrypted and stored properly?
- Are passwords secure, meaning do they contain uppercase and lowercase letters, numbers, and special characters, and

are they at least eight characters in length? Are passwords being changed on a periodic basis?

- Are you using multifactor authentication everywhere possible? MFA requires two or more verification factors before you are given access to an application, online account, etc. For example, one verification is a username and password and another factor could be an authentication application code or a code texted to your cellphone.
- It is not enough that all PHI coming in and out of your organization is HIPAA compliant. If there is an app for your business, it too needs to be in compliance with HIPAA rules.
- If you use paper forms, how are you storing, protecting and/ or destroying these forms?
- When you talk about patient PHI data, are you making sure that others do not overhear that information?
- Do you use an email or web-filtering service, which would provide an additional layer of protection against cybercriminals? Again, this is not a specific HIPAA requirement – it is a best practice.
- Another is central log management, a service that records security events, such as when employees log on and off their computer, an excellent idea from a security standpoint and crucial when used to find a bad actor if a breach occurs.

In my 20 years in IT and IT risk management, I have seen all too often, especially in smaller offices, a reluctance to get on top of HIPAA compliance and take proper and full measures to safeguard their PHI. They do not understand the ramifications of being HIPAA compliant and why they should do it.

There are literally tens of thousands of reasons to do so. Just ask the North Carolina–based dental practice about the $50,000 civil monetary penalty imposed on them after a patient submitted a complaint to the Office of Civil Rights alleging an unauthorized disclosure of their protected health information in response to a negative online review of the practice. Or the $62,500 fine levied against an Alabama dental practice whose owner and operator ran for state senator, prompting his campaign manager to send letters to all of his patients informing them of his candidacy.

In assessing penalties for HIPAA violations, there are four categories, according to *HIPAA Journal*:

- **Tier 1:** A violation the covered entity was unaware of and could not have realistically avoided, and they had taken a reasonable amount of care to abide by HIPAA rules. Fines per violation range from $100 to $50,000.
- **Tier 2:** A violation that the covered entity should have been aware of but could not have avoided, even with a reasonable amount of care (but falling short of willful neglect of HIPAA rules). Fines per violation range from $1,000 to $50,000.
- **Tier 3:** A violation suffered as a direct result of willful neglect of HIPAA rules in cases where an attempt has been made to correct the violation. Fines per violation range from $10,000 to $50,000.
- **Tier 4:** A violation of HIPAA rules constituting willful neglect and where no attempt has been made to correct the violation. The minimum fine is $50,000 per violation.

I like the idea of using an auditor or company that can verify that a company is HIPAA compliant, someone who will tell you that you have a problem that must be fixed, and then verify and attest that you are doing it correctly.

I have worked with auditors who have told me that, in many cases, the fines are so extreme that smaller businesses cannot cover them and are forced to close. All because they have not been doing what they needed to do regarding PHI.

Perhaps my biggest piece of advice to business owners is that they need to get onboard if they want to promulgate a culture of compliance within their company. If they don't, the rest of the company will not follow. It is especially important in the post-pandemic 'new normal,' when many workers are working at home. In those more relaxed and familiar surroundings, it is easy to forget about the compliance piece of the puzzle to ensure there is still protection. When people worked exclusively in the office, physical data protections were easier to enforce than at home.

For example, people tend to write information down on sticky notes or pieces of paper. Office protocol may call for those to be shredded. That is hard to enforce when someone is working at home.

HIPAA may seem daunting at first glance, but there are companies (like mine, for example) that help companies with compliance by assisting in compliance readiness and performing specific training for HIPAA as well as cyber security, which will help prevent data breaches and protect your company against bad actors trying to get to it.

Security is evolving frequently, and it is necessary to understand the changing risks and adapt in an ever-changing cyber security landscape. For example, our company utilizes managed detection and response software instead of just security event and incident management software to keep up with monitoring for potential security breaches.

The penalties for not being HIPAA compliant are too great for companies to risk. To let it slide and not establish a culture of compliance within your business…Inconceivable!

About Craig

Craig Peterson is the founder of Digital DataComm, which has offices in Orem, Utah, and Colorado Springs, Colorado. This year, 2022, marks the company's 20th year.

Craig is a Utah native whose passion for computers began in elementary school and grew through junior high and high school. In college, he focused on networking and security. Desktop support started him on his career path. He spent more than 20 years in the corporate world running larger IT functions.

The dot-com bust in the early 2000s compelled Craig to start Digital DataComm. While he continued to work in the corporate world, he also did IT consulting (which afforded him the opportunity to buy "really cool" IT tools, he jokes), but he saw the need among small and medium-size businesses to focus on cyber security and compliance such as HIPAA and PCI DSS. From Managed Services Provider (MSP), he expanded Digital DataComm to become a Managed Security Services Provider (MSSP). The company is in the process of becoming SOC 2 and HIPAA certified.

Health care businesses are a special focus for Craig Peterson. He noticed smaller businesses do not have the time or resources to fully understand HIPAA's complexities and the things they need to do to protect themselves from compliance violations and security breaches. All too often he has seen small business owners put their cousin or brother in charge and treat security like it is a side project.

An Eagle Scout himself, Craig supports his local BSA troop. He is also a certified gun instructor with the NRA and a certified Red Cross instructor. His varied hobbies (besides computers, of course) include cars, camping, fishing, hunting, sport shooting, hiking, waterskiing, wakeboarding, and snowboarding. When he is not outdoors, he enjoys reading suspense novels. He lives in Provo with his wife. They have six children and one grandchild.

Digital DataComm's mission is to WOW customers by providing proactive services, simplifying, and securing technology, and creating an awesome support experience. Craig wants his clients to know they can rely on him as a dependable and trustworthy expert resource.

For more information, contact Digital DataComm:
- Email: info@ddcit.com
- Web: https://www.ddcit.com/
- Phone: 801-356-9333

CHAPTER 7

WHY IT'S A DANGEROUS MISTAKE TO DUMP THE ENTIRE BURDEN OF COMPLIANCE ON YOUR INTERNAL IT DEPARTMENT

BY STEVE SEROSHEK,
President, Avaunt Technologies

With each passing year, compliance regulations are getting increasingly demanding. There is the need for more reporting, more monitoring, more security policies, more data-loss prevention measures, and more control arising over data sharing. Of course, these new compliance requirements involve more resources – namely, time and financial expense.

However, there is, after all, a reason. Just look at the headlines today, and you'll see a tidal wave of new cybercrime. Ransomware attacks with the attackers demanding absurd amounts of money. Data breaches that expose customers' private data or patients' records. Cybercrime is becoming a lot more frequent and causing considerably more interruption and damage.

81

Even though there are more regulations, more scrutiny, and far more risk today, too many organizations, including medical institutions, are dumping the entire burden of compliance on their internal IT departments. This is a major mistake that will most likely cost them in the long run.

Here are the five most important reasons why forcing your IT team to bear the brunt of compliance is a dangerous mistake:

Reason #1: The IT Team Is Not the Policy Team

All too often, hospitals and clinics tend to dump Health Insurance Portability and Accountability Act (HIPAA) compliance entirely onto the IT department. While IT certainly plays a major role in setting and implementing security standards, they shouldn't shoulder the entire burden of compliance. There has to be some connection between the people side of HR and the tech side of IT.

First, there must be a compliance officer. This position is typically not in IT because IT personnel are not policy creators. Rather, technicians are very linear-focused and analytical. Therefore, it's challenging for technicians to write logical policies and procedures that will keep their organization in compliance.

Second, in order for a compliance plan to be effective, it requires buy-in from everyone throughout the organization. That requires the department that's accustomed to dealing with people throughout different departments. That requires HR.

However, because effective compliance is rooted in a proven cyber security plan, IT should absolutely be involved from the beginning. An effective compliance strategy involves a collaboration between IT and HR with constant communication. That combination of HR and IT experts is what Avaunt Technologies brings to your organization.

Reason #2: Compliance Measures Are Not Properly Enforced

HIPAA mandates that someone within your organization oversee enforcing the procedures. Therefore, someone with training in compliance from HR should be well-placed to manage and enforce compliance.

The first order of business is determining who within your organization is going to enforce the compliance rules. That depends on the management structure of the environment. IT can't always enforce company-wide training.

Next, you need a compliance officer, who generally comes from HR. Typically, internal IT departments aren't knowledgeable about HIPAA. They don't know the rules about reporting and the metrics behind it. Besides, HIPAA compliance involves divulging patient records, which must be kept in the strictest of confidence.

Proper compliance enforcement throughout your organization doesn't come by ruling with an iron fist. It's about relationship-building. That's why when our team at Avaunt Technologies manages compliance, we build that relationship with the IT side, human resources, as well as everyone within your organization who plays a role in compliance.

We begin by getting information about your network, where your security stands and where you can improve. We closely analyze your current HIPAA state and security stack to find any weaknesses. From the beginning, we make it clear that we're not there to take over your IT. Rather, we become an extra resource for you, to complement your IT needs while ensuring you stay in compliance.

Reason #3: Introduces Natural Biases to Give You a False Perspective of Your Compliance

If you asked your internal IT team to tell you just how protected your network is from a ransomware attack, would they tell you the truth? If you asked them if all the compliance boxes are checked, would they be forthright and honest? Do they even know what boxes to check?

Oftentimes, when your in-house IT department is responsible for implementing compliance measures, it can be a fox-guarding-the-henhouse mentality. Because they set up the security, they have a natural bias to believe they have all security precautions in place. Whether they're unsure of their technology environment or simply presenting an overly optimistic view of security, a compliance audit will eventually reveal the realistic truth.

Over Avaunt's 30-year history of working closely with internal IT teams, we have discovered that IT managers sometimes have a narrow outlook on their own infrastructure, making them less likely to consider alternate and improved security strategies.

Additionally, we frequently act as advisors and consult with internal IT teams. Because we bring a fresh set of eyes to your network, security, and compliance needs, we often introduce solutions that have eluded techs who have been there for 10 or 15 years. While veteran technicians can certainly be protective of what they've built, sometimes it is helpful to have a review from a third party.

Our highly trained team comes in with a holistic approach, and this ensures that everything is evaluated and scrutinized. It also warrants that we deliver the reality we have found without bias with relation to people, products, or processes. The end result is an IT management and security solution that meets the ever-evolving needs of the organization and keeps cybercriminals at bay, all the while satisfying strict compliance requirements.

Reason #4: Management Can Negatively Influence Compliance Standards

No matter how structured you think your organization is in

adopting compliance standards, the process can easily be derailed because of human nature. Most businesses have a hierarchy of employees, starting with the leadership and reaching all the way down to the administrative team. While upper management often dictates compliance protocols throughout the organization, they also have the power to influence compliancy.

Case in point: In the medical industry, doctors typically run the show. Oftentimes, the doctor is the partner. Therefore, when a doctor speaks, people on his team are quick to respond. When they want something done, it typically gets done.

Unfortunately, it's not uncommon for doctors to expect a different set of rules for themselves as well as their peers. They may go to their internal IT department and say something like, "I really hate having to log in every time I use my computer. I'm wasting valuable time trying to figure out my password. Then it locks me out. Can you make an exception for me?" It's a classic case of someone with influence not wanting to comply with protocols set up and observed by everyone else on the team.

Because that doctor or partner is typically highly influential, IT often relaxes the rules a bit for upper management. This preferential treatment creates multiple problems. First, when other team members discover that their bosses don't follow the same rules and expectations, it can negatively affect morale, office relations and even productivity. Second, the level of security is greatly reduced, which leaves their organization highly vulnerable. Third, they would most likely fail their next compliance audit, which could bring serious consequences, such as penalties, fines and, even worse, a damaged reputation.

When you team up with Avaunt Technologies to secure your network and ensure your organization is in compliance, we strip out all influences to put everyone on the correct path to success. From the start, we build your IT and compliance the right way. When everyone within your organization follows the same rules, it's far easier to adhere to all compliance guidelines, keep your business in good standing and avoid heavily consequential security risks and hefty fines.

Reason #5: Internal IT Teams May Procrastinate or Be Too Busy to Do It Right

There is no doubt that maintaining adequate compliance year after year is a pretty daunting task. That's because it must involve several team members from both IT and HR, doesn't provide black-and-white instructions and can take a considerable amount of time to complete. Because tasks that are seemingly difficult and can cost considerable time and money, many organizations choose to ignore them.

Just as many Americans put off doing their taxes until the week they are due, people tend to procrastinate when it comes to reporting for compliance. They put it off until December and then unsystematically crank it out just to show what they did for the year, without knowing the level of accuracy.

Too often, team leaders underestimate the importance of following compliance protocols. These are completely disregarded, allowing the team leaders time to mistakenly focus on other priorities until the day they're involved in a costly breach. This results in them being reprimanded and forced to pay a fine or get a warning, and that is when they finally realize the consequences of not following these compliance protocols.

Aside from your IT team putting off compliance reports until the last minute, do they really have the necessary time and/or skill to allocate to this important undertaking? Even if they did, would you want them to take their eyes off of making sure your computers, network and cyber security are all functioning as they should?

At Avaunt Technologies, we know IT and compliance are equally important to preserve your security and protect your customer or patient records. Therefore, we take the time and employ the manpower to always make your compliance a top priority.

THE COST OF NONCOMPLIANCE OR EVEN UNDERCOMPLIANCE

Every year, businesses in the US are forced to devote serious time and lots of effort – and even lose a good chunk of their revenue – to taxes. While nobody wants to go through the hassle and expense of paying taxes, it's required by law. Sure, you can keep quite a bit of your money and save a lot of staff time by not paying taxes. But ignoring taxes can result in penalties, fines, a possible audit and even jail time!

Just like taxes, compliance is a massive undertaking that can eat away considerable staff time and even cost money to manage. That's why far too many organizations today choose to ignore compliance regulations, spend a minimum of effort on it or wait until the last minute to get it done. It's why they abdicate responsibility by putting compliance on the backs of your inexperienced IT technicians.

However, the cost of neglecting compliance can be steep. In fact, the cost of noncompliance is estimated to be over three times higher than the cost of compliance. Businesses lose about $1,500,000 on average due to a single noncompliance event.

But it's far more than just money lost. It can most likely mean damaging the reputation of your business, which could cost you considerable customers and patients and ultimately shut your doors.

Recently I did a quick search on a website for the Office of Civil Rights, which is part of the US Department of Health and Human Services. By law, medical facilities are required to report any breach where patient records were compromised. In just seconds, I found that in January 2021, a major hospital in South Carolina suffered an attack where a hacker used email as the vector to breach 189,761 patient records.

That means that, by law, all 189,761 patients had to be notified that their private information – possibly including their credit card numbers and patient history – was exposed. Once those exposed patients receive that email from the hospital, what is the likelihood they will get medical treatment from that same medical facility next time?

Beyond the lost customers and reputation fallout, that hospital was looking at massive fines. With a maximum fine of $1,500,000 per year, that kind of money could have more than covered ensuring their complete compliance, which could have prevented this cyber-attack.

A DIFFERENT APPROACH TO THINKING ABOUT COMPLIANCE

Business owners tend to think of compliance and reporting as a nuisance. They approach it as simply government-regulated busywork that consumes time and costs money but serves little purpose.

We at Avaunt see it differently. Compliance is a blueprint to keep your data and patient records out of the hands of cybercriminals. It helps protect you from paying out hundreds of thousands of dollars in ransomware. While it may feel like a burden at the time, it's one of today's most proven ways of avoiding the fines and penalties, lawsuits, regulatory scrutiny, reputation, and customer loss, and even imprisonment that may follow a breach. Of course, that's not even mentioning the massive ransom payments and possible weeks of downtime.

It is with confidence that we say we know your IT team is doing a commendable job at keeping your organization running and productive while protecting your data from a malicious cyber-attack. That's their expertise. But are they experts when it comes to writing, reporting, managing, and enforcing compliance?

That IS our specialty. Not only can our IT experts work alongside your IT department to maximize efficiencies and minimize risks, Avaunt Technologies can ensure your organization has all the 'green lights' and boxes checked when it comes to maintaining compliance.

HOW TO FIND THE RIGHT IT PROVIDER FOR YOUR CYBER SECURITY AND COMPLIANCE NEEDS

When you're looking for an IT provider to offer cyber security and compliance direction, service is top of mind. However, when every IT provider claims to have exceptional customer service, you need to ask the right questions to get to the real truth. At a minimum, I suggest asking any prospective IT provider these questions:

Can you provide proof of your quality customer service?

Never simply accept an IT provider's word that they have great customer service. Rather, ask them to show you their client feedback reports. If they don't have a client feedback system, they may be exaggerating their level of support. Our customer feedback system at Avaunt Technologies provides real-time metrics showing what our clients think of us. We are proud to share our high customer service ratings with anyone.

Do you offer after-hours support, and if so, what is the guaranteed response time?

Some IT providers must think that small businesses only have technology problems on weekdays from 9:00 a.m. to 5:00 p.m. That's because those are the only hours they may take your call. Hackers also know this. It's why Saturdays and holiday weekends are when many cybercriminals infiltrate networks. That way they are free to steal as much data as they can before employees notice they've been breached the following Monday morning.

At Avaunt Technologies, we respect your time. Therefore, we strive to eliminate as much of your downtime as possible. We answer our phone live so you're immediately connected to a technician.

In addition, we pride ourselves on our 15-minute response time. We know that time is money. That's why we promise never to waste yours. From the moment you call with an IT issue, we immediately address your concern. Then, in less than 60 minutes, your dedicated technician is assigned. In most cases, we can fully resolve your issue in less than a day!

Is your help desk local or outsourced?

There are few things more frustrating than wasting countless minutes on hold to speak to an IT help desk. Then, when they finally do answer, there's a massive communications block because English is not their native language. To make matters worse, they don't know your history as a client or can't fully resolve your problem.

That's what happens when IT providers go the cheap route and outsource their help desk services. Fortunately, at Avaunt Technologies, we provide a dedicated technician to your account who will get to know you and your company, as well as your preferences and history. When you work with a local help desk technician, they'll be more capable of successfully resolving your IT issues and handling things the way you want.

Perhaps our extreme focus on customer service, responsiveness and dedicated technicians is the reason why several of our clients have been with us for over two decades. If you are in need of compliance or IT services, visit www.avaunt.com for more information.

About Steve

"Find what you love doing, and you'll never
have to work a day in your life."
~ Mark Twain

Steve Seroshek has enjoyed the privilege of never having to work, because every day he gets to live his passion. From the age of 12, Steve loved tinkering with electronics and computers. He carried his fanaticism with technology to Washington State University, where he earned an electrical engineering degree. After college, Steve spent years as a leader in Microsoft's Windows and handheld-device initiatives.

While Steve appreciated the technology focus of companies he worked for, he always knew he could provide a higher degree of service to Washington small businesses. In 2002, he was finally able to fulfill his dream as a business owner when he purchased Avaunt Technologies.

"I like to think of Avaunt Technologies as a Swiss Army knife of IT. We have to be experienced in multiple areas of IT," Steve says. As a seasoned managed services provider (MSP), Avaunt is divided into three areas of focus: managed services, a private hosted cloud service and project-based services. Among the several dozen industries Avaunt serves are health care, manufacturing, legal, finance, as well as the nonprofit arena.

At Avaunt Technologies, Steve has adopted a "service first" approach to IT and cyber security. He regularly reviews technician feedback scores to ensure everyone on his team is exceeding customers' needs. That's why Avaunt maintains an intense focus on response time, follow-up, and one-on-one interactions.

When asked how his MSP has thrived in the competitive Washington area for over two decades, Steve flashes a smile and says, "We take great pride in helping local organizations grow. Many clients have been with us for over 20 years. And because a lot of our staff have been with Avaunt for over 10 years, they're like family to our clients."

From a one-man business back in 2002, when he first acquired Avaunt Technologies, Steve Seroshek now has two locations. Most of his clients are primarily located in the Pacific Northwest, with some scattered throughout the United States. His company is regularly requested to perform HIPAA compliance and network security audits for health care practices.

If you are in need of compliance or IT services, visit:
- https://www.avaunt.com

CHAPTER 8

THE E-6 PROCESS FOR HIPAA COMPLIANCE
THE SIX PIECES EVERY COMPLIANCE PLAN MUST INCLUDE

BY RUSTY GOODWIN,
Executive Consultant – The Mid-State Group

While handling compliance issues over the past two decades, I've discovered that you must understand the risks of noncompliance and factor in the human component.

'Governance,' 'risk' and 'compliance' are terms that get thrown around a lot. We define governance as anything we make ourselves do. Compliance is anything someone else makes us do. Risk is what we are trying to avoid or mitigate by using governance or compliance.

Over the years, companies nationwide have had to train their employees about Occupational Safety and Health Administration compliance and human resources compliance. After decades of compliance, it's hard for business owners to even imagine noncompliance with OSHA and the various HR regulations. Even Health Insurance Portability and Accountability Act compliance is not a brand-new concept. But this whole idea of cyber compliance for HIPAA can seem overwhelming for a lot of organizations right now. Add to that the requirements many insurance carriers are now embedding in their contracts to get a cyber liability insurance policy, and IT departments and C-suite executives are finding they have a lot piled onto their plates. It can be hard to wrap their arms around all that is expected of them.

The good news is, you can rein in the human component and get people to cooperate when you have the right process in place. I remember the beginning of the movement to make cars safer with seat belts. When the government made seat belts the law of the land, many, many people resisted. They didn't like being told what to do. Even though a lot of lives were being lost, people still wouldn't comply and refused to wear seat belts. To combat this resistance, education campaigns were launched, statistics were shared and context was given about why seat belts are important. Today, it is a rare exception to find someone who will not wear a seat belt. Instead, most people self-govern and put them on voluntarily. It's become a habit. The culture was changed through context and education.

A few years back it became apparent our company needed to pay more attention to cyber compliance. In the process of tightening up our own compliance posture, we discovered our clients needed help with it as well. While OSHA, HR and HIPAA compliance may have different requirements or technical parts, there is one challenge that is the same: *convincing people to govern themselves.*

The importance of this cannot be understated. No matter how much money you spend on technology, there is nothing the IT department can do to protect you from yourself unless employees cooperate. Even if you spend a fortune to keep your company secure, it only takes one human error to unravel everything. Much can be done to improve your risk posture, but it will require understanding and improving the human component. Globally, we spend a quarter of a trillion dollars annually to secure our hardware and software, even though over 90% of all breaches are caused by humans. How do we adopt the type of behaviors that can help mitigate this risk? That's what the E-6 process is for – to develop compliance as a culture and help protect you from the human component of compliance.

WHAT IS THE E-6 PROCESS?

The E-6 process is the easy-to-remember, repeatable method I use to help clients with the human component in their companies so they can successfully create self-governance and better practice HIPAA compliance. It's a method you can use to wash, rinse and repeat. (I recommend repeating the entire process at least once a year.)

WHY IS THE E-6 PROCESS IMPORTANT?

There is a mistake that organizations everywhere make when trying to get their employees onboard with HIPAA compliance. They start with the WHAT instead of the WHY. Author Simon Sinek would not approve! You must start with the WHY! It begins when organizations purchase stacks of hardware or software to help solve their problems without knowing how to use it or why they are buying it, other than they were told to do so. If you don't know why you're buying it or how to use it, then technology really doesn't do you any good. The E-6 process tells you the why, the how and the what.

The order of these steps is what I've found to be most successful. It's important to note that if you miss a step or go out of order, you may find the process won't be as successful.

Step 1: Evaluate

You can't know where you're going unless you know what your

starting point is. Imagine trying to put in an address for your GPS, but the GPS doesn't know where you are. It can't possibly tell you how to get to your destination without that critical information.

Similarly, you must have a good assessment of where you are starting. Assessing your hardware and software is extremely important. However, you also want to evaluate the human component. For example…

- Do you conduct phishing training?
- Do you do cyber security awareness training?
- Do you have access control for who can see what data?
- How do you protect your data?
- What are your HR policies around cyber hygiene?

There are many more questions to ask, but this is a good start.

You also want to evaluate what you want your culture of compliance to look like and how much disruption your business can tolerate in the process of getting there.

For example, let's say you are working in a doctor's office with patient data. You might require employees to log in every time they try to open any application that has protected health information (PHI) or work with a new patient. You may even require multifactor authentication. On the flip side, a facility that sells medical supplies might only need to log in once each day.

Obviously, the first example is much more disruptive. Ask yourself, what kind of disruption can my business tolerate? How much is required for my organization to be compliant?

The compliance journey is different for everybody. You must evaluate; otherwise, you don't really know where to start. Plus, determining what best practices will be appropriate for your organization will help you with your compliance requirements. You may be able to document why a control point is too disruptive for your facility. For instance, if a PA or an RN had to change their password for every patient seen, it would be difficult to get anything done. Therefore, the medical center will document what their specific protocol is for password management and why.

There are all kinds of differences you're going to come across, but related to HIPAA, and in recent years to insurance carriers, what matters is that after evaluating your risk, you need to:

1) have the appropriate policies, procedures, and best practices in place
2) train for, and continuously improve on, those policies and procedures
3) document EVERYTHING. If it is not documented, it did not happen.

Step 2: Educate

Once you've assessed the situation and know where you are going, the next step is to gather the rest of your organization and tell them *why* compliance matters. Unfortunately, this is where a lot of companies fall short. They give employees training. They give them policy manuals to read when they are onboarded. They give them all sorts of information and knowledge, but they stop short of telling them *why* it's important. It's not enough to give your employees all the knowledge they need to be compliant. You must tell them WHY they should care about this knowledge before they will truly commit to your success. In short, you're telling them, "Here is what will happen to the company (and maybe even to you) if we blow it, and here's what we're going to put in place to give us the surest path to compliance success."

Give examples!

Too many times, I see HIPAA training that just gives out a list of dos and don'ts without the WHY behind them. Or the training simply says, "You can't give out anybody's personal information." But how does this happen and why? Typically, an employee who is out of compliance is not a bad actor. They just don't understand the best practices and the reasons behind them. It could be they are careless. It could be they aren't thinking. Some examples to include are:

- Employees speaking about patient information so loudly on the phone that others can hear.
- Paperwork not concealed properly, so prying eyes can see it.
- Sharing more than the minimum information necessary with providers and staff during the patient's care.
- At one facility, an employee made a social media post asking people to pray for a particular patient, but neither the patient nor the patient's family was ready for that information to be public.
- Sometimes, there may be employees who have access to PHI or ePHI (electronic protected health information) who should not have access.

All these issues can cause a lot of trouble not only for the organization but for the individual employee as well.

Let your employees know who you are protecting.

It's not just your organization and the patients they will be protecting. Ultimately, your employees will be protecting themselves.

HIPAA is an unusual beast. In the event of a breach of data or privacy, it's not just the organization that can be held liable. The individual(s) involved can also be held personally liable. HIPAA fines range between $50,000 and $1.9 million depending on the severity, and in cases of criminal breach, the fines can come with jail time as well. Employees should care about this.

Other reasons include negative – and even nasty – publicity and the damage it can do to your reputation when you are breached.

Plus, the cost of breach notification alone averages $150 per name and can be even higher, depending on the facility in question. When a company is breached, the company is required to notify every name in the database. Recently, I asked a prospect what it would cost at $150 per name to notify everyone if the database was breached. After doing some quick calculations, it was determined it would cost $3.7 million. When I asked if that would be a bankrupting event, the prospect said it would. Insurance would only cover a tiny fraction of that amount. When everyone understands the context behind compliance, it helps the CFO better understand the need to spend the money to ensure the organization practices compliance. And the right WHY helps the employees understand that their noncompliance could cause the company to go bankrupt, resulting in no more paychecks.

In addition to fines or penalties for noncompliance, in recent years, executives and board members have been found personally liable for breaches under the idea of duty of care and fiduciary care. Lower-level employees may not be concerned with HIPAA or government regulations, but it is important to help them understand that not being compliant puts your organization, your clients' data and even your employees' data at risk. Often that same computer system or network that houses all the customer data also houses all the employees' data. Their pay rates, retirement plans and social security numbers are all types of information your employees don't want stolen. But this is rarely mentioned to them. Let your employees know that by helping protect the company, they are helping to protect themselves.

The WHY behind it, when you give people the proper context, will make it easier for them to care about the WHAT you just imparted to them.

Step 3: Equip

Now that you've evaluated and educated them on why they ought to care, you must give them the tools they need to be successful. Start by telling your employees that now that they understand why compliance is important, you're going to give them the tools they need in order to be successful with compliance.

It's crucial to give them EVERYTHING they need. Your internal IT department or your managed services provider (MSP) will typically have a suite of security products in place. Make sure you include things such as phishing simulation training so employees know what red flags to look for; bringing in an expert on cyber security awareness to teach your employees what things they should pay attention to; helping them understand proper password management. You'll also want to have your physical security team come in to show what's important to protect in different areas of your facility.

Imagine asking your employees to drive a nail without a hammer. They understand why the nails are important. They know what the nails are used for and where the nails should go, but you don't give them a hammer. When you tell them to go drive the nails, they aren't going to be able to get the job done. Similarly, you must invest in – and give employees – the necessary tools so they can pass the compliance test and meet all your expectations.

(It's important to note that the tools are going to be different depending on your situation. To fully understand what tools you'll need on the technology side, I recommend you read the other chapters in this book where this will be discussed further and also contact a professional MSP that specializes in HIPAA compliance.)

This equip phase must be a joint effort between leadership, the IT department and physical security. The employees need to know they have *everything* – all the necessary tools – to pass the test you're getting ready to give them when asking them to comply with HIPAA.

Step 4: Empower

Now they have context and understand governance vs. compliance. They have the education and tools they need. Once you provide them with all that, they will feel empowered to protect themselves and the organization. When employees feel empowered, you give them every reason to do a better job. Let's look at one area as an example: phishing training.

If all you do is walk in and say, "Make sure you watch out for those fake emails," you've given them no resources to do a better job. Without any knowledge, without the why behind it, and without the tools, employees don't know what to do. What happens is:

1) they try their best but wind up opening a malicious email anyway and get discouraged.
2) they become paralyzed and are afraid to open anything, and important work doesn't get done.
3) they say, "Screw it," and click on everything.

No one should feel powerless. When you give employees permission to do a better job, they almost always will do a better job. From my experience, I can tell you almost 100% of the time when we introduce phishing simulation training into an organization, the first time an MSP does a campaign, at least half the people or more will click on a bad email. Over time, with consistent training, the click rate gets down to only 1% or 2% at worst, and often a 0% click rate, because now we've given people the knowledge and the tools they need to recognize malicious emails. What starts happening is not only do employees recognize the malicious email, but they also start reporting it. This is just one example of empowering employees to do a better job.

Step 5: Engage

Employee engagement is vital to your success. The good news is that once you've gone through the first four steps, you'll be amazed at how engaged your employees are. Engagement means compliance is no longer falling on one department's shoulders. It's not just the IT department or the HR department or management working on the process, you have the *entire* team of people working toward compliance. Everyone is engaged in protecting each other, protecting their data and protecting their facilities, instead of the entire burden falling just on IT.

Here's why engagement matters so much. When employees become engaged, everything improves. Gallup statistics show that employee engagement positively impacts your organization and is an important predictor of your company's performance, even in a tough economy. According to Gallup.com, "Gallup researchers studied the differences in performance between engaged and actively disengaged work units and found that those scoring in the top half of employee engagement nearly doubled their odds of success compared with those in the bottom half." Companies with engaged employees outperformed bottom-quartile companies by 10% in customer rates, 22% in profitability and 21% in productivity. Plus, the top quartile also had lower turnover, lower absenteeism, fewer safety and patient safety incidents, and fewer quality defects. (Source: Gallup.com)

Step 6: Encourage

The last step is to keep encouraging your employees. When they are improving, tell them so and let them know they are doing a great job. For example, you might say, "When we first started phishing simulations, you were clicking on bad emails about once a month. However, I've seen that you haven't clicked on a single bad email in six months! You're doing a great job of keeping our organization safe."

For the employees who don't have it mastered, encourage them to watch training videos or refresh them about what they need to know and remind them why it's important. Here, you might say, "I've been reviewing the reports and you seem to be struggling with identifying phishing emails. Remember that when you click on a bad email, you're putting yourself at risk because your personal data, including your Social Security number, is housed on our computer network. Watch these training videos and they will help you get better at identifying bad emails."

In addition to what the government requires, insurance carriers are now requiring a much more robust risk posture when companies apply for cyber liability insurance. Included in this list of must-haves are multifactor authentication, password management, phishing training, cyber security training and proof you are successfully backing up your data. All things that will require you to govern yourself appropriately.

The E-6 process will not only help you comply with HIPAA and satisfy your insurance carriers, but it will also ultimately create a culture of compliance in your organization. Once you have that, all the rest becomes so much simpler.

About Rusty

Rusty Goodwin has been helping companies with success, compliance, growth, and risk management for over two decades. As an organizational efficiency consultant at the Mid-State Group, he specializes in organizational efficiency, process streamlining, resource optimization and compliance. His expertise is in creating simple and powerful solutions so busy leaders can devote the time and attention necessary to carry them out, and continue them as successful, ongoing habits. Given the nickname 'the Fixer' by his clients, Rusty gained the moniker due to his innate ability to introduce immediate fixes that streamline processes and increase the bottom line. By bringing order to chaos and making seemingly impossible changes, he continually helps organizations find the answers.

After years of helping organizations create a culture of compliance regarding their Occupational Safety and Health Administration Compliance and Human Resources Compliance, Rusty began working as a consultant for managed services providers (MSPs) that hire him to help translate and simplify cyber-compliance risks and issues for their clients. With a lifetime certification in DISC Profiles/Five Behaviors and a certification in PXT Select™ (an assessment measuring thinking styles, behavioral traits, and interests), he assists by assessing the need and breaking things down simply for his clients. This enables his clients to understand all the risks and issues surrounding either the Health Insurance Portability and Accountability Act compliance framework or the Cybersecurity Maturity Model Certification framework, those implications, and the interventions they can make. (When clients take a good assessment of their risk profile, Rusty finds that most of the time they choose to govern themselves even more conservatively than what's required of them through compliance. Therefore, he is driven to help clients understand the risk and context around compliance.)

Rusty also specializes in comprehensive organizational audits, and can provide, or assist with, compliance assessment, HR/safety integration assessment, data breach strategy and response, employee education and training, and framework best practices.

His mission is to change the attitudes and culture surrounding compliance so that every company strives to self-govern and develop a culture where compliance is the norm, not the outlier. He does this by helping C-suite executives wrap their heads around cyber-compliance, what it means, and why it's important for them to take compliance into their own hands by self-governing as well as understanding all the risks relating to compliance and noncompliance.

Rusty serves on the Product Advisory Council for Kaseya, a best-in-breed technology company that assists IT professionals in efficient management, security, and backup IT under a single pane of glass. He is also working with Technology Marketing Toolkit to develop a model for MSPs that will help their clients with compliance.

Connect with Rusty at:
- Email: rgoodwin@themidstategroup.com.

CHAPTER 9

WHY HIPAA ISN'T JUST FOR U.S. COMPANIES... IT'S ALSO FOR CANADIAN COMPANIES

BY GARETH MCKEE,
Founder & CEO, Burnt Orange Solutions

A friend of mine, who is also Canadian, lived in Vermont for almost four years. During his time in the United States, he concluded that Americans rarely think about Canada. One reason is that Canada is seldom mentioned in the newscasts, in US print and online journalism or in casual conversation. The opposite is true regarding how Canadians approach the United States. Many Canadians follow US politics more closely than they follow Canadian politics. Canadians have access to many US TV channels, which they devour. While there are some Canadian channels available in the US, they have limited market saturation.

It's a similar situation when it comes to HIPAA and Canadian medical businesses. Most people in the health care industry in Canada are familiar with the term 'HIPAA', but if you asked them for the name of the Canadian equivalent, they'd be hard-pressed to come up with an answer. They know HIPAA, which stands for Health Insurance Portability and Accountability Act, from seeing it in the news, on medical-related websites and from talking to their associates. But they also know that because they're governed by the laws of Canada, HIPAA is not something they have to worry about.

In the United States, HIPAA awareness has increased to where the leadership of most covered entities (CEs) know it's something they must take seriously. A Managed Services Provider (MSP) sales representative will call up someone at a CE and attempt to book an appointment to talk about HIPAA compliance. American MSP websites prominently advertise that they specialize in HIPAA compliance because they know it's something their target audience is aware they need. HIPAA compliance companies market aggressively. Because businesses have been fined, had audits done and been dragged into court due to complaints or data breaches, Americans are much more aware of it. And they are more fearful of the ramifications of ignoring it.

"I DON'T CARE"

It's a different situation in Canada. While they are aware of HIPAA, when I mention that there is a Canadian equivalent, Canadians don't know anything about it. Nor do they want to. When I tell them we have privacy laws in place in Canada and they must make sure their data is encrypted, secured, backed up and so on, they don't seem to care. They have no idea whether they're compliant or even what being compliant entails. They're not worried about it at all. They just see it as more unnecessary government red tape.

So, while in the United States, MSPs will phone up a prospective client and try to get an appointment to talk about HIPAA compliance, I've tried that up here with the Canadian equivalent of HIPAA, which is called PIPEDA (Personal Information

Protection and Electronic Documents Act), and it doesn't work. People aren't aware of it, and they don't think it applies to them. I don't advertise compliance in any of our literature or on our website. People simply aren't interested. I've had conversations with doctors about our privacy laws and the need to adhere to them, and their response is simply, "I don't care."

AWARENESS AND RAMIFICATIONS

So, why don't Canadian businesses take compliance as seriously as they should? There are two main reasons, in my opinion:

The first is awareness. In the United States, HIPAA is a big topic of conversation. When there's a data breach or someone gets fined, it's all over the news. In Canada, if a data breach occurs, it might make the news once, but then you hear nothing else about it.

The second reason is that the perception of the ramifications is different. In the United States, people are aware and concerned that if they don't become HIPAA compliant, they could be fined and embarrassed, and their reputation might take an enormous hit. In Canada, PIPEDA-delinquent companies are rarely taken to court.

$100,000 PER VIOLATION

The Office of the Privacy Commissioner of Canada (OPC) has a flowchart on its website (https://www.priv.gc.ca/biens-assets/compliance-framework/en/index#) that outlines how a PIPEDA complaint or issue is handled. The goal seems to be to find a resolution and avoid taking the company to court. (If it gets to court, you've most likely done something extremely stupid or something with malicious intent.) The fines can be quite steep – up to $100,000 per violation. If your case goes to federal court, you can be legally ordered to (a) comply with PIPEDA asap, (b) publish a notice when you're compliant, and (c) pay damages to the person who brought the complaint.

Three offenses are deemed criminal:

1) Intentionally destroying information after someone requests access to it.
2) Retaliating against an employee who complained about a breach or refused to breach PIPEDA.
3) Obstructing officials investigating a complaint. But, as mentioned, most businesses seem to believe they'll never be the target of a PIPEDA audit and be forced to pay a fine of any magnitude.

Having said that, Canadian CEs (and *every* business that collects, uses, and discloses customer information) need to be more aware of the importance of keeping their clients' data safe and how a data breach or ransomware attack can potentially cripple them financially, cost them clients and ruin their reputation.

A COSTLY BELIEF

One incident I have firsthand knowledge about involved a rental company. My company, Burnt Orange Solutions, does the IT and security for a business owned by a man who also owns 50% of a rental company. The owner of the other 50% of the rental company, however, doesn't believe in Internet security. He believes that viruses are created by the virus company to keep people buying antivirus software. They got hit with a ransomware attack a few years ago. They lost all their data and didn't have a backup. They had tens of thousands of SKUs. And they had hundreds of items rented out and no way of determining who had all that equipment. They lost most of the rented-out equipment and had to rebuild their database from scratch. They couldn't do business for months. They lost millions of dollars. To add insult to injury, they paid the $18,000 in ransomware, but the cybercriminal never sent them the decryption keys as promised. I do not know if they were fined by the OPC.

So, what exactly is PIPEDA and why should you care? To answer that question (and many others), let's first look at HIPAA.

WHAT IS HIPAA?

HIPAA became law in the United States in 1996. It's a series of regulatory standards that oversee the lawful use of protected health information. HIPAA compliance is enforced by the United States Office of Civil Rights and regulated by Health and Human Services. HIPAA applies to CEs (doctors, clinics, psychologists, dentists, chiropractors, nursing homes, pharmacies) and business associates, which are businesses a CE engages to help it carry out its health care activities and functions.

PIPEDA: THE CANADIAN VERSION OF HIPAA

PIPEDA became law on April 13, 2000, to promote consumer trust in e-commerce, and its implementation occurred in three stages. Starting in 2001, the law applied to federally regulated industries (such as airlines, banking, and broadcasting). In 2002, it was expanded to include the health sector. On January 1, 2004, PIPEDA came fully into force. On that date, any organization that collects personal information during commercial activity was covered by PIPEDA, except in provinces that have similar privacy laws. (Seven provinces have privacy laws that have been declared by the federal Governor in Council to be substantially similar to PIPEDA. More about this in a minute.)

PIPEDA's mission is to make sure that organizations are responsible and accountable for protecting all data collected, regardless of the province or industry. It mandates that organizations are transparent when they collect information. They must explain why it's being collected and how it will be stored. PIPEDA also gives individuals the right to privacy over their information.

The following is an excerpt from the OPC website (https://www.priv.gc.ca/en/) that outlines an organization's responsibilities for each of the ten PIPEDA principles:

- **Principle 1 – Accountability.** An organization is responsible for personal information under its control. It must appoint someone to be accountable for its compliance with these fair information principles.
- **Principle 2 – Identifying Purposes**. The purposes for which the personal information is being collected must be identified by the organization before or at the time of collection.
- **Principle 3 – Consent.** The knowledge and consent of the individual are required for the collection, use, or disclosure of personal information, except where inappropriate.
- **Principle 4 – Limiting Collection.** The collection of personal information must be limited to that which is needed for the purposes identified by the organization. The information must be collected by fair and lawful means.
- **Principle 5 – Limiting Use, Disclosure, and Retention.** Unless the individual consents otherwise or it is required by law, personal information can only be used or disclosed for the purposes for which it was collected. Personal information must only be kept as long as required to serve those purposes.
- **Principle 6 – Accuracy.** Personal information must be as accurate, complete, and up to date as possible to properly satisfy the purposes for which it is to be used.
- **Principle 7 – Safeguards.** Personal information must be protected by appropriate security relative to the sensitivity of the information.
- **Principle 8 – Openness.** An organization must make detailed information about its policies and practices relating to the management of personal information publicly and readily available.
- **Principle 9 – Individual Access.** Upon request, an individual must be informed of the existence, use, and disclosure of their personal information and be given access

to that information. An individual shall be able to challenge the accuracy and completeness of the information and have it amended as appropriate.

- **Principle 10 – Challenging Compliance.** An individual shall be able to challenge an organization's compliance with the above principles. Their challenge should be addressed to the person accountable for the organization's compliance with PIPEDA, usually their Chief Privacy Officer.

HOW DOES HIPAA DIFFER FROM PIPEDA?

HIPAA's primary concern is to protect health insurance information. PIPEDA's focus is much broader. It focuses on all types of personal data, including health information. However, all private-sector commercial businesses in Canada that collect, use, and disclose personal information must be PIPEDA compliant— *not just health care, but ALL related businesses.*

In the United States, HIPAA is a federal law, as PIPEDA is in Canada. However, in Canada, some provinces have their own laws, rules, and regulations regarding the gathering of this data.

HOW THE PROVINCES AND TERRITORIES HANDLE DATA SECURITY

Three provinces are exempt from PIPEDA because they are similar to PIPEDA in that they don't just deal with health information, but with all personal data. They are (i) British Columbia, where the Personal Information Protection Act came into effect on January 1, 2004; (ii) Alberta, where their version of PIPA also came into effect on January 1, 2004; and (iii) Quebec, which has the Private Sector Act.

In addition, four provinces have passed legislation that deals strictly with health information: Ontario has the Personal Health Information Protection Act, which came into force on

November 1, 2004; New Brunswick has the Personal Health Information Privacy and Access Act (PHIPPA), which came into effect on June 19, 2009; Newfoundland and Labrador have the Personal Health Information Act, which was passed in 2008; and Nova Scotia has the Personal Health Information Act, which came into force June 1, 2013.

In 2012, in Saskatchewan (where I live), someone found thousands of medical records in a Regina dumpster. The privacy commissioner called it "the worst breach of patient information" his office had ever seen. In response, the government added amendments to HIPA, which was originally brought in as law in 1999. Even so, HIPA takes a backseat to PIPEDA in most cases in Saskatchewan.

Prince Edward Island, Manitoba, and the three territories (Yukon, Northwest, and Nunavut) do not have their own HIPAA-style law, so PIPEDA applies.

Note: British Columbia and Nova Scotia are the only two provinces that require patient health information not to leave the province even if it's encrypted. When data is transferred out of the country, it becomes subject to the laws where the data is stored.

Which law must the provinces that have a health care privacy act follow?

The answer is both laws. Here is a question and answer from a document put out by the New Brunswick government that addresses this issue specifically.

I am a health-care professional already covered by PIPEDA. Does the provincial legislation, PHIPAA, now replace the federal legislation, PIPEDA?

Since Jan. 1, 2004, organizations in New Brunswick that collect, use, or disclose personal information, including personal health information in the course of "commercial activities" such as private physicians' offices, private healthcare clinics and laboratories, and pharmacies have been subject to the federal Personal Information Protection and Electronic Documents Act (PIPEDA). PIPEDA has been identified by health-sector stakeholders as especially problematic for the organizations that collect, use, or disclose personal health information for healthcare purposes since it was not developed with the special needs of health care in mind. PHIPAA provides more detailed rules than PIPEDA and also provides some additional flexibility in privacy practices for the health sector.

With the introduction of PHIPAA, custodians engaged in commercial activities will continue to be bound by PIPEDA. However, they will now also be required to comply with PHIPAA with respect to the personal health information that they collect, use, disclose and maintain.

While there is generally more flexibility under PHIPAA with respect to the use and sharing of personal health information with other health-care practitioners, there are additional obligations imposed on custodians under the Act.

ARE MOST CANADIAN COMPANIES PIPEDA COMPLIANT?

Here are some results from a 2019–2020 survey of Canadian businesses published on the OPC website (https://www.priv.gc.ca/en/opc-actions-and-decisions/research/explore-privacy-research/2020/por_2019-20_bus/):

- Sixty-two percent (62%) of companies have designated someone to be responsible for privacy issues and the personal information that their company holds (up from 59% in 2017 and 57% when tracking began in 2011).
- Six in ten (60%) companies have procedures in place for

responding to customer requests for access to their personal information (up from 47% in 2017).

- More than half (58%) have procedures in place for dealing with complaints from customers who have concerns about how their information has been handled (up from 51% in 2017 and 48% in 2011 when tracking began).
- Fifty-five percent (55%) have developed and documented internal policies for staff that address privacy obligations under the law (up from 50% in 2017).
- Four in ten (39%) regularly provide staff with privacy training and education.
- Large companies (i.e., companies with at least 100 employees) are more likely to have put in place a series of privacy practices, to have policies or procedures in place to assess privacy risks and to have a privacy policy.

While the above might be accurate, in my experience, I would say that most Canadian companies are not PIPEDA compliant because they're simply not aware that PIPEDA exists. Many IT people assume their data is fine, and as long as it's backed up and they can recover it, they're happy because their business is functioning. They don't know what their PIPEDA responsibilities are. We often go into a new client and find their desktop computers and laptops are not even password-protected. A few years ago, a client bought six laptops and they wouldn't pay the extra $50 to encrypt their hard drives. The first weekend they had them, the laptops were stolen from their office. Each laptop had a treasure trove of client information – names, addresses, banking details and so on – that was now accessible to the person who stole them.

IN CONCLUSION...

If you are handling health care–related information, PIPEDA, the Canadian equivalent of HIPAA (or the provincial equivalent, depending upon your home province), applies to you. *It's important to highlight that every business in*

Canada that collects, uses, and discloses personal information while conducting commercial business must also be PIPEDA compliant. You must designate a senior person within your organization to take responsibility for PIPEDA compliance. You need to develop clear policies and procedures to make sure you follow the 10 principles of PIPEDA, and you must keep records of the information you gather, the consent given and how you use and share information. You must also be sure that everyone who accesses information in your business does so without breaching the PIPEDA guidelines. If you don't, you are not only breaking the law, but you are also putting your company, your employees, and your patients/customers at risk.

I also recommend that you enlist a third-party business, like my Burnt Orange Solutions, to help you become PIPEDA compliant. This way, you won't have to worry about the technical aspects of it or risk being fined because you missed something. You'll have peace of mind knowing that everything you do in your business complies with the rules and guidelines laid out by the provincial and/or federal government and that your patient/customer data is safe.

About Gareth

Gareth McKee is the founder and CEO of the award-winning Canadian MSP Burnt Orange Solutions, in Saskatoon, Saskatchewan. Burnt Orange is the only IT support firm in Saskatoon specializing in cyber security, computer, and network support with guaranteed response times, and focusing particularly on small and medium businesses. Gareth and the Burnt Solution team also specialize in keeping businesses compliant with both Saskatchewan's Health Insurance Protections Act (HIPA) and Canada's Personal Information Protection and Electronic Documents Act (PIPEDA), which keep your business and health care data safe from ransomware and cyber-attacks.

Gareth comes from a family who have served in the British Army and the Royal Navy. In 1992, Gareth joined the Royal Corps of Signals, the communications brand of the British Armed Forces. In 1997, he was injured when he jumped from an aircraft and the landing didn't go as planned. With time on his hands during his rehabilitation, his innate knack for technology and security blossomed. Gareth soon became the person who made sure all the computer systems worked properly and were secure from foreign cyber-invaders. His work laid the groundwork for the next generation of battle planners to strategize using innovative technology.

With the skills and knowledge to provide all-encompassing IT protection, Gareth founded Burnt Orange Solutions in 2008 in the United Kingdom, to provide hardworking businesses with the same comprehensive protection. In 2012, Gareth and his wife, Alison, a physician, moved Burnt Orange Solutions to Saskatoon. For Gareth and Alison, Saskatchewan has proven to be a great place to raise their two sons, and they are now proud Canadian citizens.

Burnt Orange Solutions offers comprehensive IT services, from cloud backup and network security to data backup/restoration and PIPEDA and HIPA compliance for a simple, monthly rate. This proactive approach to IT makes their clients' budget processes much easier and maximizes their uptime. Besides health care, Burnt Orange Solutions also services the construction and financial industries.

Gareth and the Burnt Orange team share a passion for Benjamin Franklin's famous quote "Don't put off until tomorrow what you can do today." They believe that if there's one more thing you can do at the end of the day that will make tomorrow better, it's worth doing. Gareth's goal for his team is to connect on a personal level and build long-term relationships of trust with every client. Burnt Orange does this by upholding these five guarantees daily: 1) provide the service and quality promised, 2) complete the job on time, 3) charge the price quoted with no surprises, 4) communicate honestly and be responsive to customer needs, and 5) resolve any issues with customer satisfaction in mind.

Burnt Orange Solutions was awarded the 2019 Consumer Choice Award, the Best Business of 2018, the 2017 MSPmentor Award and CRN's Next Generation 250 Winner in 2018 and 2019. They've also been given the thumbs-up by TrustSaskatoon.com, a website that helps consumers and businesses find companies that operate with integrity.

For more information, contact Gareth at Burnt Orange Solutions:
- Email: garethmckee@burntorangesolutions.com
- Phone: 306-986-2600
- Web: https://www.burntorangesolutions.com

CHAPTER 10

WHAT'S YOUR CYBER-HYGIENE SCORE?

BY SHANE SERRANT,
Managing Partner, Alternative Systems

Ever since we were old enough to step on a stool and stand over the bathroom sink, most of us learned how to brush our teeth twice a day. And our parents likely adopted daily routines to support our oral hygiene as soon as we cut those first incisors. Habits that lead to good oral health have been embedded into our minds through repetition over years, and if we use them consistently, we have a decent shot at preventing tooth decay and gum disease. Also, if we are lucky enough to have access to regular checkups, cavities are likely to be detected early and filled to stave off further damage.

Alternative Systems' goal as a managed services provider (MSP) is to help your organization adopt a culture in which cyber-hygiene is at the forefront, so it feels like second nature – just like oral hygiene. This is the best method to keep the data on your computer network safe from attacks. It requires continual maintenance, including regularly updating anti-malware software, apps, browsers and operating systems, installing firewalls, two-factor authentication, security assessments and network leak patches.

Our approach also includes an educational component so we can make informed choices about our actions. Just as we teach youth about chewing tobacco's link to mouth cancer in hopes of preventing them from using it,[1] anyone who uses the network needs to know how phishing works and the destruction it can cause, so they know to abstain from clicking on suspicious links.

Any reputable practitioner encourages maintenance, just as a dentist would never tell you your teeth and gums are so healthy at this moment in time that you can stop brushing and flossing, and you'll never need another cleaning or X-ray. And for all you password-change resisters out there, think of what would happen if you used the same toothbrush for many months or even years. The bristles would fray every which way and leave big gaps, making it impossible to thoroughly clean your teeth. The same is true for a stale password that's been around forever. Regularly changing passwords is a necessary step to keep access points clean and prevent harmful forces from entering.

What does your cyber-hygiene score mean?
A cyber-hygiene score represents your network's likelihood of thwarting off a hack and its potential severity. Regular system maintenance can keep this number high and your system at less risk of a malware attack or other breaches, and a watchful eye improves the chances that potential threats will be detected quickly.[2]

How is a cyber hygiene score calculated?
The most important thing to keep in mind is that a cyber-hygiene score is not a fixed number. It can change from day to day, based on how diligently an organization implements habits to maintain good health.

A consistent method for all organizations to analyze vulnerabilities and assess interventions is the Common Vulnerability Scoring

1. https://www.cancercenter.com/cancer-types/oral-cancer/risk-factors
2. What All Businesses Should Know About Cyber Hygiene (https://sopa.tulane.edu/blog/cyber-hygiene)

System.[3] The base section of the score considers attack vector (network, adjacent, local, or physical), attack complexity, privileges required, user interaction required, changes in scope, as well as confidentiality, integrity, and availability. Ratings of low, medium, high, and critical help determine necessary security management tactics.

Similarly, credit scores are used to indicate borrower risk levels and can be sustained by positive behaviors such as regularly paying bills on time and using a small percentage of available credit, which can bring up the number and open opportunities for growth through future lenders. A lower number, however, may lead to higher interest rates or even denial of credit.

FACTORS WITH HEAVIER IMPACT

Make sure you're aware of which factors have the most weight with cyberscores. It's not much different than credit score formulas, which are comprised of payment history at 40%, credit history and usage at 20%, followed by credit checks and available credit.[4]

Passwords have a much greater impact on security than data inventory. This is because you use passwords every day, so your contact points are high. To improve security, create passphrases that are difficult to crack and relatively easy to remember, such as MySonWillTurn5ThisYear, then save them in a password manager. Or use a random password generator.

You only need a few strong passwords and to create new ones every few months or, at a minimum, once a year. Try doing this on or near a holiday to begin to associate this simple act with memorable dates. We help reinforce this with our clients by

---

3. Mell, P., Kent, K., and Romanosky, S. (2006), Common Vulnerability Scoring System, IEEE Security & Privacy [online], https://tsapps.nist.gov/publication/get_pdf.cfm?pub_id=50899; and https://www.nist.gov/publications/common-vulnerability-scoring-system

4. VantageScore® 3.0., according to Chase Credit Journey

sending reminders that the Fourth of July is coming up or New Year's Eve is around the corner.

We'll continue to talk about passwords until we stop finding them written on sticky notes – in plain sight – at clients' medical offices! We confiscate every single one, and the user has to reset them. We've seen extraordinary things, like patients reaching over the desk to steal passwords. It's not worth the risk.

KEY ACTIONS TO KEEP YOUR CYBER-HYGIENE SCORE HIGH

Most of the recommendations on the good cyber-hygiene list are already contained within the Health Insurance Portability and Accountability Act of 1996 (HIPAA) guidelines required by covered entities – health care providers and clearinghouses, health plans and business associates. Here are some of the ways you can keep your network healthy:

- **Regularly update security software, browsers, and operating systems to the latest versions to help prevent hackers from easily breaking into your network.**
  In addition to patching the server and desktops, remember to have your IT department or MSP check for gaping holes in the firewall, switches, wireless access points and other equipment, and patch where needed.

- **Stay informed about which devices are used on your network.**
  Make sure an individual or team monitors all of them for vulnerabilities.

- **Regularly train all staff on cyber security hazards and safety measures.**
  We notice medical clients frequently leave a lot of confidential patient information out in the open. Part of training is informing staff of dangers, then showing them

how to clean things up. It all goes back to the dental hygiene conversation.

They need to be aware of how frequently people enter a medical practice and attempt to access another patient's information. In one situation, a patient plugged a flash drive into a computer to try to get data. Turns out, he brought malware into the network. Fortunately, since the office had already hired our MSP, we were monitoring the system remotely and were able to alert the doctor that we detected an unauthorized access attempt and stopped the damage.

- **It may seem rudimentary, but train staff to lock their machines when stepping away from the desk.**
  If someone accesses the computer, your organization could be liable for the harm an intruder (often disguised as a patient) inflicts if they have accessed private info because it was readily available. It's not unlike the case of a Texas mom who had to pay for 31 McDonald's cheeseburgers when her two-year-old placed an order with *DoorDash* using her unlocked smartphone. May this precocious-toddler story be a reminder of how easily one can access a device and make transactions using your identity.

Many situations are more nefarious in nature. It is far too common that someone will go to a medical office specifically to attempt to find out an address or other details about an estranged boyfriend, girlfriend, spouse, or kids. For example, a husband separated from his wife may visit the family pediatrician to find where she's living with their children. He may go under the guise that he's there to pay a bill, but his intentions can be quite harmful.

Why do they target the physician's office? Medical information is more than likely up-to-date, so it's accurate. Every time you go to the doctor, they want to know if your address has changed and if you have the same insurance.

Creeper exes aren't the only ones who know this. Hackers do too. And this personal medical information is a gold mine.

In order to protect your patients, it's imperative to train staff to never give out information. Always ask the requester to verify what they already know. You'd be surprised at what people try! It's sad. We educate about these types of scenarios because we encounter them often.

- **Create, implement, monitor, and enforce policies.**
  Take policies and procedures seriously. Make sure you have at least the following policies in place:

  - **IT Acceptable Use Policy.** This provides a list of things employees are allowed to do using company computers and equipment, as well as restrictions.
  - **Business Associate Agreement.** A HIPAA requirement for all health care professionals, this sets the guidelines for anyone who may have access to patient health information. Practices with physical files must hold cleaning crews or others who enter the space to these rules.
  - **Breach Notification Policy.** This provides the steps for what will happen should a breach occur, including who will be contacted, what message will be released and how patients will be notified and protected.
  - **Employee Confidentiality Agreement.** A document that all employees must sign upon hire to legally bind them to keep patient and company information confidential.
  - **Information Privacy Policy.** This states that your organization and associates will keep patient information safe and private and will not release them to anyone without permission.
  - **Right of Access Policy.** This outlines that patients actually own their health records kept by the practice and states that you will provide them access to their health information when needed.

- **Document everything.**
  Like all of cyber-hygiene, documentation is an ongoing process. As you update processes and equipment, such as switching to a new payroll system or upgrading a credit card processor, document it so you are protected. Then reassess security.

- **Embed healthy cyber-hygiene practices into your organization's culture.**
  This includes engaging in conversations with employees about habits on a regular basis. While technology is at the root of cyber security, it's important to emphasize relationship-building. An MSP or internal IT department is only able make improvements if they pay attention to what employees need – what frustrates them, what's not working.

  We ask to be invited to clients' quarterly meetings to learn about any changes to the organization and any new concerns. It makes a difference because our interest and presence make them feel comfortable sharing updates, which makes us better equipped to help them.

- **Know what equipment you have and where it is located.**
  Keep inventory up-to-date. Otherwise, you may not notice if something has gone missing. If it's lost, and we are your MSP, we are legally obligated to report it. If it contained patient information, the consequences to you will be high.

  For medical practices, make sure machines are not capable of leaving exam rooms. This can be solved by locking them in a cage. Since doctors today have moved to laptops or tablets for patient notes, make sure to teach them to form the habit of never leaving patients alone with their devices.

  Dental practices in particular have a bit of a unique setup in that all the patient care is done in a single room. Staff need to be trained to make sure the computer is positioned so it cannot be accessed easily and is locked when the dentist, dental assistant or hygienist exits.

- **Limit employee access privileges.**
  HIPAA rules are designed to ensure that each person has the least amount of access needed to do their job. This helps your practice stay compliant and mitigates risk.
- **Install surveillance cameras where appropriate.**
  Use cameras to help track unauthorized access of patient information, with the exception of medical exam rooms, for privacy reasons.

  If you run a dental office, install a camera and turn it on any time a patient is under sedation. When people wake up, they may think anything that happened in their dream really occurred. This can be a legal disaster. Claims of abuse while a patient is asleep are very common.

Just as your dentist would rather not have to perform a root canal, as an MSP, we don't want to have to remediate breaches. We prefer that your organization is safe and secure. We also want to help our clients adhere to compliance to ensure that cyber insurance will pay them for damages in the event of a hack. Insurance companies continue to get hit hard paying out cyber insurance claims, so they are making the requirements for getting a policy more stringent. Make sure you are actually following the rules listed on the policy application – otherwise you will likely get a claim denial when an incident happens.

And if your office is too small to have a designated human resources person or IT manager to facilitate training, we recommend outsourcing this task to experts such as an MSP. You may think we only serve technical concerns, but we in particular also lead clients in compliance training.

It's important for health care practice leaders to be well-versed in all the facets of cyber-hygiene maintenance – and view it as a team sport. Especially considering the increasing requirements by insurance companies for health care organizations, it's more crucial than ever to cover all the bases to ensure compliance. And the more your staff is involved in ongoing training, conversations, and daily habits, the higher your score and the healthier your organization will be overall.

About Shane

Shane Serrant set out to become a high school teacher, earning dual degrees in secondary education and electrical engineering. After working in the classroom for one semester, though, he accepted a job in information technology as a technical support engineer. He went on to found IT company Alternative Systems in 2007, where his strengths as an educator continue to shine. Shane's approachable methods of communicating with clients and explaining technical topics influence company culture, and his team emulates his style.

Perhaps the most instrumental experience in defining the company was his tenure at a large nonprofit organization (NPO) where he designed custom programs that enabled patients to apply for medical insurance grants more efficiently than ever before. Back then, paper forms and long approval periods were the norm, so his work revolutionized the submission process. Moving to the digital realm allowed patients to receive funding – and essential medical treatment – faster.

Shane and his team dug deeper and learned that most recipients did not have checking accounts. They were paid by check, and to cash it they relied on services that charged such a high percentage that they would fall $20 to $50 short of the insurance premium cost. To solve this problem, Shane developed a system to add the money to a debit card, which eliminated fees and sped up the timeline.

While Shane was happy to use his talents to help many people get medical care, when the NPO's annual revenue jumped from $115 million to $250 million, he recognized the disparity between NPOs that could afford to build digital tools and smaller ones with tighter budgets that could only offer antiquated systems. This prompted Shane's decision to find a way to provide true corporate-level access to local businesses that previously couldn't afford it. A few years later, he started Alternative Systems, and by 2010, the company took off.

At this time, cyber security was becoming more of a necessity. And as health care practices increasingly became targets, Shane's company chose to focus on protecting patient information. They established the mission to use technical expertise to help patients and deliver high-level, enterprise-grade IT services at a fraction of the typical cost. Today, they serve many independent health care practices and are the exclusively preferred vendor for Maryland State Dental Association members.

Their goal is to make sure IT clients understand what they're doing and why. Shane strives to make messages connect with the subject and prides himself on making the content interesting enough that he doesn't put the doctors to sleep.

Originally from Brooklyn, New York, Shane studied at the University of Maryland–College Park. He stayed in the DC area, where he lives with his wife and three very active daughters. Shane's office is boldly decorated with signed posters of New York Yankees and Giants players. He always wears Yankee gear for checkups with his Red Sox–fan dentist, who laughs in response, pointing to the tools at his disposal that can cause pain.

Contact info:

Alternative Systems
16701 Melford Blvd, Suite 400
Bowie, MD 20715

- Web: https://www.alternativesys.com/
- Email: info@alternativesys.com
- Phone: 301-799-3000